# Statistics in Education

# Arnold Applications of Statistics Series

*Series Editor:* **BRIAN EVERITT**
*Department of Biostatistics and Computing, Institute of Psychiatry, London, UK*

This series offers titles which cover the statistical methodology most relevant to particular subject matters. Readers will be assumed to have a basic grasp of the topics covered in most general introductory statistics courses and texts, thus enabling the authors of the books in the series to concentrate on those techniques of most importance in the discipline under discussion. Although not introductory, most publications in the series are applied rather than highly technical, and all contain many detailed examples.

# Statistics in Education

**Ian Plewis**
*Institute of Education*
*University of London, London, UK*

A Member of the Hodder Headline Group
LONDON • NEW YORK • SYDNEY • AUCKLAND

Copublished in North, Central and South America
by John Wiley & Sons Inc.
New York • Toronto

First published in Great Britain 1997 by Arnold,
a member of the Hodder Headline Group,
338 Euston Road, London NW1 3BH
http://www.arnoldpublishers.com

Copublished in North, Central and South America by
John Wiley & Sons Inc., 605 Third Avenue,
New York, NY 10158-0012

© 1997 Ian Plewis

All rights reserved. No part of this publication may be reproduced or transmitted
in any form or by any means, electronically or mechanically, including
photocopying, recording or any information storage or retrieval system,
without either prior permission in writing by the Copyright Licensing Agency:
90 Tottenham Court Road, London W1P 9HE

*British Library Cataloguing in Publication Data*
A catalogue record for this book is available from the British Library

*Library of Congress Cataloging-in-Publication Data*
A catalog record for this book is available from the Library of Congress

ISBN 0 340 64628 4
ISBN 0 471 19487 5 (Wiley)

Publisher: Nicki Dennis
Production Editor: Wendy Rooke
Production Controller: Rose James
Cover Design: M2

Typeset in 10/11 Times by AFS Imagesetters Ltd, Glasgow
Printed and bound in Great Britain by J W Arrowsmith Ltd, Bristol

# Contents

**Preface** ix

**Description of Datasets** xi

**1 Statistics and Educational Research** 1
   1.1 Introduction 1
   1.2 Statistical models 1
   1.3 Types of educational data 5
   1.4 Educational research designs 6
   1.5 Statistical inference 8
   1.6 Plan of the book 9

**2 Building the Foundations: Modelling Continuous Responses with Linear Regression** 11
   2.1 Introduction 11
   2.2 Simple regression with a continuous explanatory variable 11
   2.3 Transforming variables 16
   2.4 Simple regression with a categorical explanatory variable 18
   2.5 Multiple regression with continuous explanatory variables 20
   2.6 Multiple regression with categorical explanatory variables 24
   2.7 Multiple regression with a mixture of categorical and continuous explanatory variables 26
   2.8 The assumptions of regression 28
   2.9 Other topics in regression 29
   Exercises 31

**3 Populations with Structure: Multilevel Models for Continuous Responses** 33
   3.1 Introduction 33
   3.2 The hierarchical nature of educational data 33
   3.3 Partitioning variability by level 35

vi  *Contents*

|  |  |  |
|---|---|---|
| 3.4 | Multilevel models for school and teacher effectiveness | 38 |
| 3.5 | Random intercepts and random slopes | 42 |
| 3.6 | Explaining variability in intercepts and slopes | 47 |
| 3.7 | Multilevel models for intervention studies | 50 |
| 3.8 | Software for multilevel modelling | 53 |
|  | Exercises | 53 |

**4 Growing and Changing: Repeated Measures of Continuous Responses** — 55
- 4.1 Introduction — 55
- 4.2 The hierarchical nature of repeated measures data — 55
- 4.3 Basic multilevel growth curve models — 59
- 4.4 Explaining variability in growth — 64
- 4.5 Complex variation at level one — 67
- 4.6 Missing data — 69
- 4.7 Intepretational issues — 71
- Exercises — 74

**5 Two by Two Tables and Beyond: Modelling Binary Responses** — 77
- 5.1 Introduction — 77
- 5.2 The two by two table — 77
- 5.3 Larger contingency tables — 80
- 5.4 Introducing logistic regression — 81
- 5.5 Logistic regression with more than one explanatory variable — 84
- 5.6 Methods for dependent samples — 88
- 5.7 Concluding remarks — 91
- Exercises — 91

**6 Larger Contingency Tables: Modelling Categorical Responses** — 93
- 6.1 Introduction — 93
- 6.2 Larger two-way tables — 93
- 6.3 Log-linear models for modelling association — 94
- 6.4 Representing order in categorical responses — 100
- 6.5 Modelling ordered responses — 101
- 6.6 Modelling change — 103
- 6.7 Modelling agreement — 105
- 6.8 Concluding remarks — 108
- Appendix 6.1: SAS code — 108
- Exercises — 111

**7 Education as a Career: Event History Analysis** — 113
- 7.1 Introduction — 113
- 7.2 Types of event history data — 114
- 7.3 Fundamental concepts in event history analysis — 116
- 7.4 Constructing survival curves — 118
- 7.5 Statistical models for event history data in continuous time — 120
- 7.6 Statistical models for event history data in discrete time — 126
- 7.7 Models when there is more than one episode — 129
- 7.8 Conclusion — 131
- Appendix 7.1: Dictionary of terms used in event history analysis — 132

|   |   | Appendix 7.2: GLIM code | 134 |
|---|---|---|---|
|   |   | Exercises | 135 |
| 8 | **Modelling Educational Data: Further Issues** | | **137** |
|   | 8.1 | Introduction | 137 |
|   | 8.2 | Measurement errors | 137 |
|   | 8.3 | Extensions of multilevel modelling | 142 |
|   | 8.4 | Modelling structured categorical data | 145 |
|   | 8.5 | Concluding remarks | 146 |

**Answers to Selected Exercises**   **149**

**References**   **155**

**Author Index**   **159**

**Subject Index**   **161**

# Preface

I have written this book for everyone who wants to do quantitative educational research, and who is interested in using modern statistical techniques to analyse their data. I include in this group Masters students whose course requires them to do a dissertation, Doctoral students, and researchers working in universities, independent research units and in central and local government. The emphasis of the book, and its examples, is on educational research. However, the designs, methods and statistical techniques used in educational research are closely related to those used by other social scientists, and so I believe the book will be of interest to this wider audience too.

This is an intermediate rather than an introductory textbook. The focus is on statistical models, with the first half of the book concentrating on models for continuous responses and the second half on models for categorical data. I assume that readers will have done an introductory course in statistics for social scientists, covering descriptive statistics, the basics of statistical inference, something about the main statistical distributions, and, perhaps, an introduction to linear regression. I also assume that they will have had some exposure to a statistical package such as SPSS. I have kept the mathematics simple, but it is difficult to present statistical models without using some algebra, and so the ability to understand and to manipulate simple equations will make it easier to benefit from the book.

I have selected what I believe are the most useful statistical models for educational researchers. In addition, I have gone more for breadth than for depth, and have concentrated more on interpretation than on theory. Indeed, behind each of the chapters there is at least one textbook or monograph which describes the models in more detail, and I refer to these for readers who would like to go further. All the models are introduced through the medium of examples, and all these examples use real data, usually taken from recent educational research. Most of the datasets used in the book are included on the associated disk, and so readers can try to reproduce for themselves the results in the book. There are exercises at the end of each of Chapters 2 to 7 to enable the reader to try out other models on these datasets, as well to test their understanding of important principles. Answers to a subset of these exercises can be found at the end of the book.

Fitting statistical models to data requires a computer in all but the most simple cases. As SPSS is probably the statistical package most widely used by educational researchers, I have used it whenever possible. However, SPSS does not have any procedures for fitting multilevel models, and so I have used the MLn package extensively in Chapters 3 and 4. SPSS is also a little restricted when it comes to models for categorical data, and so I have used SAS and GLIM to fit some of the models in Chapters 5 to 7. However, because the use of SAS and GLIM is less widespread among educational researchers, I give in appendices the program statements I have used for those two packages.

Finally, it is a pleasure to acknowledge all the help I have received while writing this book. For many of the examples I have used data collected by past and present colleagues at the Thomas Coram Research Unit. I have gained great benefit from being part of the Multilevel Models Project team at the Institute of Education, and from the support the team received from the Economic and Social Research Council under its Analysis of Large and Complex Datasets programme. I am very grateful to friends and colleagues who found time to read and comment on draft chapters, especially Jeff Evans, Harvey Goldstein, Jane Hurry, Toby Lewis and Charlie Owen. They helped me to eliminate many errors; any that remain are entirely my responsibility, and I would be pleased to be told about them.

IAN PLEWIS<br>London<br>February 1997

# Description of Datasets

These are the datasets which are used for illustrative purposes in Chapters 2 to 7. They are also needed for some of the exercises at the end of each of these chapters. They are held as ASCII or text files in free format.

**Dataset 2.1   (DS21.DAT)**

There are 39 cases and 6 variables:

(1) School (1,2,3)
(2) Curriculum coverage
(3) Mathematics attainment, end of Year two
(4) Sex (0=boy; 1=girl)
(5) Ethnic group (0=white; 1=African Caribbean)
(6) Mathematics attainment, end of year one

**Dataset 3.1   (DS31.DAT)**

There are 777 cases and 5 variables:

(1) School (1–35)
(2) Teacher (11–352)
(3) Pupil (1–1948)
(4) Reading attainment, end of reception (mean=0, SD=1)
(5) Reading attainment, end of Year one (square root transformation)

**Dataset 3.2   (DS32.DAT)**

There are 407 cases and 6 variables:

(1) Teacher (8–47)
(2) Pupil (1–529)
(3) Curriculum coverage (mean=0, SD=1)

xii  *Description of Datasets*

(4) Curriculum coverage (raw scores)
(5) Mathematics attainment, end of Year one (mean=0, SD=1)
(6) Mathematics attainment, end of Year two

**Dataset 4.1   (DS41.DAT)**

There are 1758 cases and 8 variables:

(1) School (1–33)
(2) Pupil (1–751)
(3) Occasion (1–6)
(4) Reading attainment
(5) Zread
(6) Ethnic group (0=white; 1=African Caribbean)
(7) Sex (0=boy; 1=girl)
(8) Age (mean=0)

**Dataset 5.1   (DS51.DAT)**

There are 64 cases (which are cells here) and 5 variables:

(1) IQ (1 (low)–4 (high))
(2) College plans (0=no; 1=yes)
(3) Parental encouragement (0=low; 1=high)
(4) Socio-economic status (1 (low)–4 (high))
(5) Cell count

**Dataset 5.2   (DS52.DAT)**

There are 217 cases and 7 variables:

(1) Child ID (1–1947)
(2) Readtest (0–44; 999 is missing)
(3) Mother's education (0 (low)–2 (high))
(4) Sex (0=boy; 1=girl)
(5) Reading aloud (0–540; 999 is missing)
(6) Whether good reader (0=no; 1=yes; 9 is missing)
(7) School (1–35)

*All datasets for Chapter 6 can be easily input from the chapter tables or from Appendix 6.1.*

**Dataset 7.1   (DS71.DAT)**

There are 234 cases and 7 variables:

(1) Child ID (3003–97067)
(2) Group (0=non-returner; 1=returner)
(3) Start of episode two (months)
(4) Finish of episode two (months)

*Description of Datasets* xiii

(5) Censoring indicator (0=not censored; 1=censored)
(6) Preschool type (1=relative; 2=childminder; 4=nursery; 5=playgroup)
(7) Mother's education (0 (low)–6(high))

**Dataset 7.2  (DS72.DAT)**

There are 1384 cases and 9 variables:

(1) Response (0=no event; 1=event)
(2) Number of failures
(3) Risk set indicator
(4) Time period (1–20)
(5) Risk set size
(6) Episode start (months)
(7) Mother's education (0–6)
(8) Constant (=1)
(9) Preschool type (1–4)

**Dataset 7.3  (DS73.DAT)**

There are 1181 cases and 7 variables:

(1) Child ID (3001–99013)
(2) Episode number (1...)
(3) Start of episode (months)
(4) Finish of episode (months)
(5) Censoring indicator (0=not censored; 1=censored)
(6) Mother's education (0 (low)–6 (high))
(7) Preschool type (0–6)

# 1
# Statistics and Educational Research

## 1.1 Introduction

In very broad terms, the discipline of statistics is concerned with the collection, analysis, presentation and interpretation of quantitative data. Over the years, statisticians have made vital contributions to each of these tasks, but it is the field of data analysis which receives most attention from statisticians. Hence, a narrower definition of statistics, but one which more faithfully reflects the contents of this book, is the analysis of variability. Variation, or variability, is not only at the heart of statistics, it is also central to quantitative educational research. To give just a few examples: pupils and students vary in their rates of educational progress; teachers vary in the way they teach; schools vary in the way they group pupils; local education authorities vary in the level of resources they allocate to education. The statistician's task is, first, to measure this variability and then, much more interestingly, to try to understand, or to explain it. *Why* do some pupils get on faster at school than others; *what characteristics* distinguish less effective teachers from more effective ones; *how much* influence do historical factors have on education spending at the local level? These are the kinds of questions which educational researchers ask and which, in collaboration with statisticians, they can hope to answer. To do this we must use statistical *models*. Statistical models form the bedrock of this book, and so they are introduced in a general way at this point.

## 1.2 Statistical models

Modern statistical practice is based on statistical models. Some of these models are simple; others are complex, as we shall see in later chapters. However, at this stage, let us focus on the strengths of statistical models as conceptual tools. A statistical model aims to represent the world in a way that is accurate and relevant, while at the same time being as simple as possible. To be accurate and to be faithful to the underlying process, a model may need to be complex. On the other hand, simple models are often easier to understand and to interpret, and will sometimes be more relevant. A statistical model is rather like a map. When going directly from A to B by car, we need some basic information about the roads and landmarks between these two points. However, if we are hill walking, we need much more detailed information about paths, contours and so on. The first map can be simple, the second is necessarily more complex, but both should fit the purpose for which they are required. It is up to statisticians and substantive researchers to work together to find an appropriate representation.

There are four particular strengths of a modelling approach to data analysis. They are:

(1) a distinction is drawn between the *response* variable and the *explanatory* variables,
(2) when we use models in our analysis, we make our assumptions explicit,
(3) statistical modelling unifies techniques which, to the non-statistician, often appear not only separate but unconnected,
(4) by writing a model as an equation, or as a set of equations, and so by using the power of mathematics to simplify and to organize, we create opportunities to refine and to expand our thinking about the educational processes we are studying.

Let us first consider the distinction between the response and the explanatory variables. To start with, we should note that what is called the response or the response variable throughout this book, researchers often refer to as the dependent variable and sometimes as the outcome. Moreover, explanatory variables are often referred to as independent variables, sometimes as covariates and factors, and sometimes as predictor variables. Each of these other terms has some disadvantages, especially the term 'independent variable', because, quite simply, independent variables are not, in general, independent of each other and cannot, in any useful model, be independent of the response. Consequently, if we do not use the phrase 'independent variable', then we avoid confusion by jettisoning 'dependent variable' too.

The response is usually labelled '$y$' in any model and placed on the left-hand side of the model equation, the explanatory variables are the '$x$' variables which are placed on the right-hand side of the equation. For a lot of the processes studied quantitatively in education, it is clear, on broadly theoretical grounds, which variable is the response and which are the explanatory variables, and so we want to reflect this theory-driven distinction in our models. For example, we expect parental attitudes and behaviours to affect children's attainment at school, and so attainment, usually measured by an educational test of some kind, is the response, and the parental attitudes and behaviours are the explanatory variables. However, the distinction is not always quite so obvious. For example, we might suppose that teaching behaviours in the classroom affect how and what pupils learn, but it is possible that pupils' learning styles, perhaps acquired from previous teachers, will in turn affect, as well as be affected by, what teachers do. In these more complicated cases, we would need more complex models in which '$y$' is the response in one equation and an explanatory variable in another.

For the most part, we confine ourselves in this book to models in which the distinction between the response and the explanatory variables is clear cut. If it is not, then we enter the world of *structural equation modelling* described by, for example, Cuttance and Ecob (1987). Sometimes, we have more than one response variable and *multivariate* statistical models are then needed. Multivariate methods (such as factor analysis and cluster analysis) are not covered in this book. Instead, readers are referred to, for example, Manly (1986).

Turning to the second strength of modelling – making assumptions explicit – we clearly need some assumptions in order to move from 'noisy' and complicated real world educational processes, to the essence of these processes as represented by our statistical models. The complexity of the world is represented by our data, and we can separate these data into two components and write:

$$\text{DATA} = \text{FIT} + \text{RESIDUAL}$$
$$\text{or}$$
$$\text{DATA} = \text{SIGNAL} + \text{NOISE}$$

The better the fit, or the stronger the signal, the smaller the residual or the weaker the noise, and the closer our model comes to the real world. In other words, if we find a good fit, our explanatory variables are doing a good job in explaining, or accounting for, the variability in the response. We should, however, remember that a statistical explanation of variability is not necessarily a *causal* explanation.

We need assumptions in order to make tractable the task of finding a model which fits the data well without being too complicated. These assumptions fall into two main classes. The first class of assumptions concerns the nature of the *link* between the response and the explanatory variables. The second class consists of assumptions about the *statistical distribution* of the response. Under a narrower definition of modelling than the one used here, it is these distributional assumptions, sometimes referred to as probability models, which are the essence of statistical modelling.

Thinking first about the link, often, but by no means always, we assume that the response is linked to the explanatory variables by a linear function. At its simplest, this linear function is the straight line of Figure 1.1, representing the link between the response, $y$, and a single explanatory variable, $x$, and characterized in statistics by a *simple regression* model. In Figure 1.1, we see that a unit change in $x$ corresponds to a change in $y$ of $b$ units for all values of $x$, with $b$ positive in Figure 1.1 but not in general. However, as we shall see, we can handle many other kinds of links between the response and one or more explanatory variables.

Turning to the distribution of the response, often it is convenient to assume it has the familiar 'Normal', or bell-shaped, distribution, although again we do not always do so. The most important departures from straight line links and Normal distributions come when we model categorical variables, when the data are counts or proportions. The assumptions we make about links and statistical distributions also act to confer some unity across the different statistical approaches we use to analyse different types of data. This is our third strength of modelling.

Finally, let us briefly consider the advantages conferred by a modelling approach to statistical analysis in terms of refining and expanding our ideas,

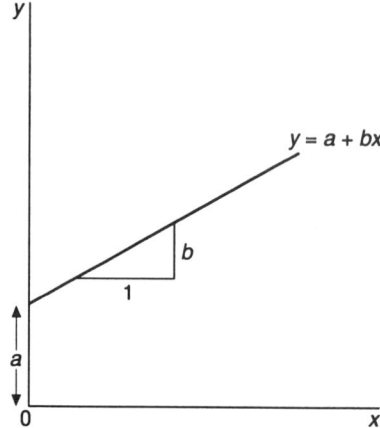

**Figure 1.1** Equation of a straight line; intercept *a* and slope *b*

hypotheses and theories about educational processes. As we shall see, the process of modelling is usually an iterative one. First, we draw on the relevant theory, policy issues and subject matter knowledge to *specify* a model. Then we fit this model to our data or, in other words, we *estimate* the model *parameters*. If we find that this first model does not fit as well as we would like, we try something different – perhaps a different link between the response and the explanatory variables, or a different assumption about the distribution of the response, or a new explanatory variable. Each of the models we fit, we subject to model *checking* – also known as model *criticism*, until we believe we have done the best we can with the data available to us. Then we give an educational or policy interpretation to our model. But the investigation need not end there; other researchers can take our data and our model, develop it, perhaps from a different theoretical perspective and with additional data, thus improving our understanding of the process.

It is sometimes argued that the procedures of statistical modelling do not give enough emphasis to descriptive and exploratory data analysis as presented by, for example, Marsh (1988). In fact, there does not need to be any conflict between the two approaches; they are essentially complementary. As we shall see, all sensible data analysis should start with an exploration of all the data, using a variety of methods, many of them pictorial, in order to know the data better, to sort out unusual, and possibly incorrect, data points known as *outliers*, and so on. But there are limits to how far description can take us. At some point we need to introduce models to give a focus to our ideas, to unify insights gained from exploratory analyses, and to provide us with a more complete understanding of the underlying processes.

It might be argued that modelling requires us to make so many assumptions as to be unrealistic. It is true that modelling cannot take place without assumptions. It is also the case that some simple statistical models are in fact simplistic. However, there is now a wide range of statistical models available to educational researchers, models that are better able to reflect the underlying educational reality. One of the most important advances in recent years has been the development of what are known as multilevel models. No longer is it necessary to ignore the fact that students are located within classrooms which, in turn, are located within schools. As we shall see in Chapters 3 and 4, multilevel models enable researchers to take account of the structure of their data in a way that was not easy to do in the past.

In this section, I have presented some basic, and very general, ideas about the importance of using statistical models when analysing educational data. These bare bones will reappear in various ways in each of the succeeding chapters, when we will put a lot more flesh on to them. Those readers who would like a more detailed, but accessible discussion of these general issues, focusing on the value of statistical models in all kinds of research, are referred to Cox and Snell (1981).

## 1.3  Types of educational data

Different research questions in education generate different kinds of quantitative data. Sometimes we want to explain, at least in a statistical sense, variation in attainment test scores, on other occasions we might be interested in why the proportion of pupils with reading problems varies from school to school. In fact, there are essentially three different scales, or levels, of measurement which we want to be able to analyse, and hence to model. We will see in later chapters how the statistical models we use vary according to the measurement scales of our response variables. We shall also see how to incorporate different kinds of explanatory variables into our models.

The first of these data scales is often referred to as *interval* or *continuous* data; we assume that a measurement on a case or unit can take any numerical value for the variable in question. An interval scale implies that differences between equally spaced numbers on the number line have the same meaning wherever they occur. In other words, the difference between 20 and 30 has the same meaning as the difference between 70 and 80. However, we do not assume that a unit with a score of say, 30, has twice as much of the characteristic represented by the variable as a unit with a score of 15. If it did, then we would have a *ratio* scale, time and money being two examples. However, ratio scales are so uncommon in education as not to be worth considering separately. Within the class of interval scale variables, there are some variations. Most of the 'raw' scores for these variables are positive whole numbers (or zero) – this is certainly so for attainment test scores – and so they are not strictly continuous. However, we often work with 'standardized' or 'z' scores where the mean is zero and the standard deviation is one, and then a range of positive and negative values is possible (although large absolute values are, of course, most unlikely). Often, and especially when we measure attainment, we assume that a measurement scale has interval properties for convenience, although there is not always a theoretical justification for doing so.

The second and third scales are categorical; the variables take just a few values, rarely more than six. These scales can be ordered or not, and the methods we use to analyse them will depend crucially on this distinction. By ordered, we mean that a higher number on the scale denotes more of the characteristic, but that the distance between adjacent scale points does not necessarily have the same meaning at all points on the scale. This is our second scale of measurement – ordered categorical data. Examples are teachers' assessments of pupils' abilities, and measures of social position. We include within this class variables sometimes described as being measured on an ordinal scale, such as ranked data.

Our third scale – unordered categorical data – is perhaps better thought of as a classification rather than as a scale; there are a small number of categories into which our units can be placed. Examples are ethnic group, school type (state funded, state aided, privately funded) and particular kinds of educational handicap. A special case of this third scale applies to variables which can take just two categories, and so, when aggregated over units, they can be expressed as proportions. These data are often known as *binary*, or as *yes/no* data. We get binary data as a result of asking certain kinds of questions in surveys (for example, 'have you passed any examinations?'), and when pupils are categorized as having or not having a particular kind of problem, for example a learning disability.

It is important to recognize that the measurement scales of many of the variables we use in educational research are not closely tied to theory, and are therefore, to some extent, arbitrary. This is especially relevant when we are thinking about our response variables, because it is the nature of the scale of the response and its statistical distribution which exert a strong influence on the method of analysis we adopt. We do not ignore scale issues for the explanatory variables, but they are less influential. We do not have the advantages of the physical and biological sciences, where most variables are measured with fixed scales – grams, centimetres and so on. To quote Duncan (1984, p.162):

> With the possible and, in any event, limited exception of economics, we have in social science no system of measurements that can be coherently described in terms of a small number of dimensions. Like physical scientists, we have thousands of 'instruments', but these instruments purport to yield measurements of thousands (not a mere a hundred or so) variables. That is, we have no system of units (much less standards for them) that, at least in principle, relates all (or almost all) of the variables to a common set of logically primitive quantities.

Because the scales are often arbitrary, we have some flexibility in the way we analyse them. For example, we can *transform* our raw data so that they come closer to satisfying the assumptions required to use a particular model. Thus, we can, for example, transform a response on an interval scale so that it is Normally distributed. However, there are a number of disadvantages with arbitrary scales, notably the fact that our results and conclusions can vary according to the assumptions we make about the scale. This is particularly so when we analyse repeated measures data, as we shall see in Chapter 4. We take up the issue of transformations in more detail in the next chapter.

## 1.4 Educational research designs

No statistical analysis can ignore the design of the study which generated the data. Before we can embark on a sensible strategy for analysis we need to know

## 1.4 Educational research designs

about sampling methods, experimental control or the lack of it, missing data and so on. This is not a book about research design. However, this section briefly describes the most prevalent designs in educational research and gives the most important references. In all the examples used in this book, we see how the data were generated, and how the research design influences the analysis.

There are two powerful types of design in educational research: *randomized experiments* and *longitudinal designs*. These two types can be combined in the same study. But, for all kinds of reasons, probably the two most frequently used design types are *observational studies* and *cross-sectional* surveys. It is usually asserted that randomized experiments produce unambiguous causal conclusions. In fact, results from randomized experiments in education are not always as clear cut as they seem. Nevertheless, the introduction of the principle of randomization into all the human and social sciences, first proposed by Fisher for agricultural research in the 1920s, has led to a number of important advances. Without randomization, we can never be sure whether observed differences on the response variable between groups have arisen because of pre-existing differences between the groups, or whether they have been caused by the variable, often called the 'treatment' variable, which differentiates the groups. Unfortunately, randomization is often difficult to implement in educational research and so a good deal of research must, perforce, rely on so-called quasi-experiments and other sorts of group comparisons, often referred to collectively as observational studies. The statisticians' task is then much more challenging, albeit perhaps more interesting, because we must try to model not only the outcome, but also the allocation process which led to units being in one group rather than in another. With randomization, this latter process is known. The classic book about experimental and quasi-experimental designs for social science research is by Cook and Campbell (1979).

The aim of an experiment in education is to induce change, often to ameliorate a problem, by introducing a treatment or intervention of some kind. However, often we are interested in understanding naturally occurring educational change or development. Longitudinal designs are then essential. With longitudinal designs, we collect data from the same observational units (pupils, teachers etc) on more than one occasion. This enables us to study both the between, or inter-unit, variation, and the within, or intra-unit, variation of a characteristic like reading attainment as it changes with age. This is illustrated by Figure 1.2 – reading attainment improves as children get older but at different rates for different children. We would like to know why some children (e.g. child A) get on faster than others (e.g. child C), and models for this question are presented in Chapter 4.

We can also use longitudinal data to estimate a postulated causal model, for example how strong the relation is between training and later earnings. Moreover, longitudinal designs create the possibility of inferring a causal model from variables which are changing over time, for example the relation between aggressive behaviour in school and exposure to media violence. Also, longitudinal designs can provide valuable data about the time, or duration, spent in a particular state such as a type of school, and about the transitions between states. Often, however, longitudinal research is deemed too expensive to conduct and the data too difficult to analyse. Instead, researchers must do the best they can with cross-sectional data, often from survey research. They are

8   *Statistics and Educational Research*

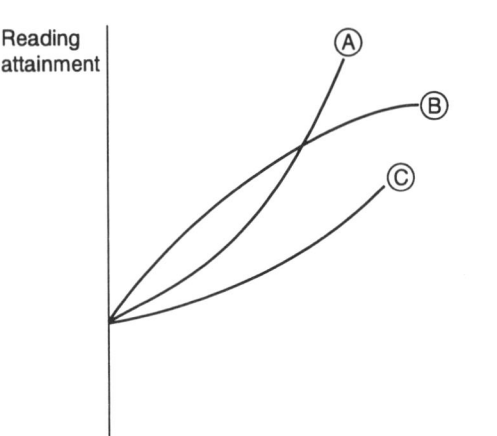

**Figure 1.2**   Reading attainment varying with age within and between children

then inevitably restricted as to the kinds of inferences about change, and the causes of change, that they can draw. Certainly, longitudinal data present their own analysis problems, especially, as we shall see, the problems created by the correlation between measurements within units and the difficulties, in educational research, of knowing exactly how to define change. Nevertheless, these problems are not insuperable. There are a number of books on the analysis of longitudinal social science data such as Plewis (1985) and the collection edited by Dale and Davies (1994).

There are other kinds of research designs which are sometimes used in educational research. For example, single subject designs and time series designs generate a particular kind of data which require particular methods of analysis. We might, for example, want to analyse A-level examination results for England for each of the last 30 years using time series methods. These methods are not covered in this book; Kratochwill (1978) is a useful reference here. Also, it is increasingly common for researchers to reanalyse data collected by other research teams. This is known as secondary analysis; it does not introduce many extra statistical problems, but it is important for secondary analysts to be familiar with the data collection and sampling strategies used in the original study. Dale *et al.* (1988) present a useful introduction to secondary analysis.

### 1.5   Statistical inference

It is important to realize that, whether longitudinal or cross-sectional, the design of a study will have an important influence on the statistical analysis, and on the generalizability of the results. Rarely do we have complete data for the population of interest – all Year two pupils in state schools in 1997 in England and Wales, for example. Instead, we have data for a sample from the population, and we hope, sometimes in vain, that this sample has been selected using some kind of probability mechanism, as described by, for example, Barnett (1991).

Techniques for statistical inference feature strongly in most introductory statistics textbooks, sometimes to the detriment of other important topics. The emphasis on inference, which is often represented solely as hypothesis testing, or tests of statistical significance, can distort statistical analysis. The upshot is that the first question researchers often ask themselves is 'what test should I use for these data?'. This leads them away from a comprehensive and imaginative analysis of their sample data, and towards a cookbook approach which will not necessarily throw any light on the processes of interest in the population. On the other hand, sampling variability cannot be ignored; results are bound to vary from sample to sample and we should never forget this.

The focus of this book is on statistical modelling of education data, rather than giving a comprehensive description of all the different ways in which we seek to make reasonable statements about what is happening in the population, on the basis of what we have found about the sample. Instead, statistical inferences are made in the context of the examples in each chapter. Sometimes tests of statistical significance are used, more often sampling variability is assessed by presenting confidence intervals for model parameters, and likelihood ratio tests are used to assess the improvements in model fit as more explanatory variables are brought into the model. Readers needing a more detailed account of statistical inference as it is used in the social sciences could consult Freedman *et al.* (1991).

## 1.6 Plan of the book

There are eight chapters altogether. Chapters 2 to 4 deal with models for continuously measured response variables. Chapter 2 introduces the fundamental statistical model – *regression* – when the response is measured on a continuous scale and there can be one or more than one explanatory variable. Chapter 3 extends the idea of regression to situations where there is more than one *level* in the data, most commonly when pupils are located within schools. Chapter 4 introduces the idea of *growth curves* for educational data, locates these ideas within a multilevel context for the analysis of repeated measures data, and pays careful attention to issues of scale and interpretation.

Chapters 5 to 7 are concerned with models for categorical responses; Chapter 5 with models for binary responses such as *logistic regression*, and Chapter 6 with models for responses with more than two categories such as *log-linear models*. Chapter 7 is an introduction to models for *event history* data. The final chapter of the book discusses outstanding issues and points the way to further, more specialized, reading.

# 2
# Building the Foundations: Modelling Continuous Responses with Linear Regression

## 2.1 Introduction

In this chapter, we consider a basic, but very important, model for analysing the variability in a continuous response. We start with the situation of a continuous response – mathematics attainment, say – related to just one continuous explanatory variable, for example curriculum coverage. We then move on to consider the case of a continuous response related to one categorical explanatory variable, ethnic group for example. Often, however, models with just one explanatory variable do not take us far enough, and a better understanding of the underlying process comes from introducing more explanatory variables, both continuous and categorical. The statistical methods we use to answer these questions come under the headings of *simple* and *multiple regression*, techniques statisticians often refer to as the *general linear model*. Models of the kind described in this chapter are fundamental. After first mastering simple and multiple regression, it is then possible to understand and use the more complicated statistical models described later in this book.

## 2.2 Simple regression with a continuous explanatory variable

Consider the following: we have a sample of pupils whose mathematics attainment is tested at the end of Year two (which is now, in England and Wales, the end of Key Stage One, when pupils are about seven years old). This is our response, $y$ (MATH). We also have a measure from their teachers of how much of the mathematics curriculum each pupil has covered during Year two. This is our explanatory variable, $x$ (CURRIC). We would expect pupils who have covered more of the curriculum during the year to have higher test scores at the end of it. We want to find out if this is indeed the case and, if so, how strong the effect is.

One way of finding out about the relation between attainment and curriculum coverage is just to calculate the *Pearson* or *product moment correlation (r)* between $y$ and $x$:

$$r = \text{covariance}(xy) / \text{SD}(y) \text{SD}(x) \quad (\text{SD} = \text{standard deviation}) \quad (2.1)$$

However, this approach has some drawbacks. The first is that correlation is a symmetric measure of association which does not distinguish between response and explanatory variable. The second is that a correlation is inversely proportional to the standard deviation of the response, so that the more variation in the response the smaller the correlation, providing the covariance stays the same. These two drawbacks mean that a correlation cannot tell us how much we would expect $y$ to change for a unit change in $x$. Hence, we prefer to use simple regression as follows:

$$\text{MATH}_i = \text{CONSTANT} + b\, \text{CURRIC}_i + \text{RESID}_i \quad (2.2)$$

or, more generally

$$y_i = a + b x_i + e_i \quad (2.3)$$

Essentially, what we have here is a slightly more sophisticated version of the equation of a straight line shown in Figure 1.1. The subscript $i$ represents the units, in this case pupils, in the sample and runs from 1 to $n$, where $n$ is the sample size. The parameter $a$ is the intercept on the $y$-axis and represents the predicted value of $y$ in the population when $x$ is zero. The parameter $b$ is the slope and indicates how much we expect, or predict, $y$ will change when $x$ changes by one unit. However, we cannot necessarily infer anything about a *causal* relationship of $x$ to $y$ just from this simple model. The $e_i$ are the *residuals* of the model, representing the noise or the 'rough' in the data arising from the fact that $x$ is an imperfect predictor of $y$. These residuals are assumed to have a mean of zero.

Both $a$ and $b$ are estimated from the sample data. There are a number of ways in which we can estimate $a$ and $b$, the most common of which is by the method of *least squares*. This means we minimize the sum of squares of the residuals, i.e. we minimize $\Sigma e_i^2$. We find that the estimated value of $b$, which we call $\hat{b}$, is:

$$\hat{b} = \text{covariance}(xy)/\text{variance}(x) = r s_y / s_x \quad (2.4)$$

where $r$ is the sample Pearson correlation, $s_y$ is the sample standard deviation of $y$ and $s_x$ is the sample standard deviation of $x$.

In addition:

$$\hat{a} = \bar{y} - \hat{b}\bar{x} \quad (2.5)$$

where $\bar{y}$ and $\bar{x}$ are the sample means of $y$ and $x$ respectively.

A proof of these results can be found in any standard statistical textbook, for example, Weisberg (1980). Under certain important assumptions, which will be discussed in the context of examples, and in Section 2.8, these least squares estimates of $a$ and $b$ have desirable properties.

Let us now expand on these ideas through an example, consisting of data on mathematics attainment and curriculum coverage from 39 pupils attending three inner London primary schools. (These data form dataset 2.1 on the disk; see p. xi) They are a subset of the data collected as part of a research project called 'Changes in the Classroom Experiences of Inner London Infant Pupils,

## 2.2 Simple regression with a continuous explanatory variable

1984–1993' (CECIL), and this study is described in detail in Plewis and Veltman (1996). The full dataset is analysed in Chapter 3. Before we consider fitting a regression line to these data, there are a number of preliminary explorations of the data which we should do.

First, we should look at the distribution of our response, mathematics attainment. We see from the histogram (Figure 2.1) that it has rather a long tail on the right, which suggests that the population distribution might be positively skewed. There are grounds for preferring our response to be Normally distributed and so we might want to consider transforming it. We do not need to worry about the shape of the distribution of $x$ but, nevertheless, it is sensible to obtain some descriptive data for the explanatory variable. Figure 2.2 shows the histogram of curriculum coverage and we see that there is a 'hole' in the distribution around the value of 250. Also, both the mean and standard deviation are rather large numbers, suggesting that a transformation might be convenient. We return to the issue of transformations in the next section.

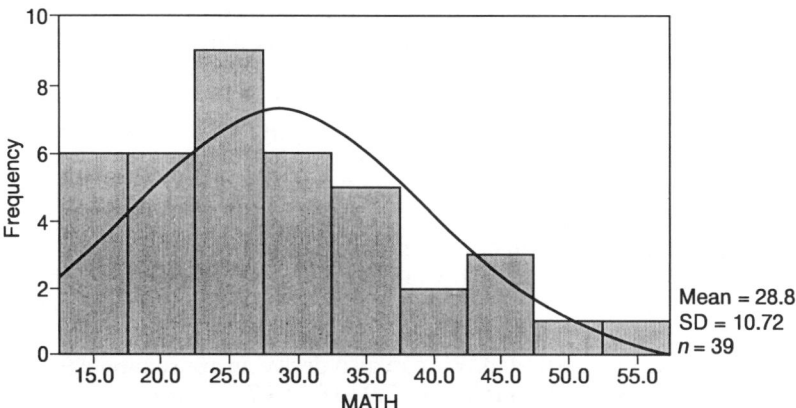

**Figure 2.1**  Histogram of mathematics attainment, end of Year two

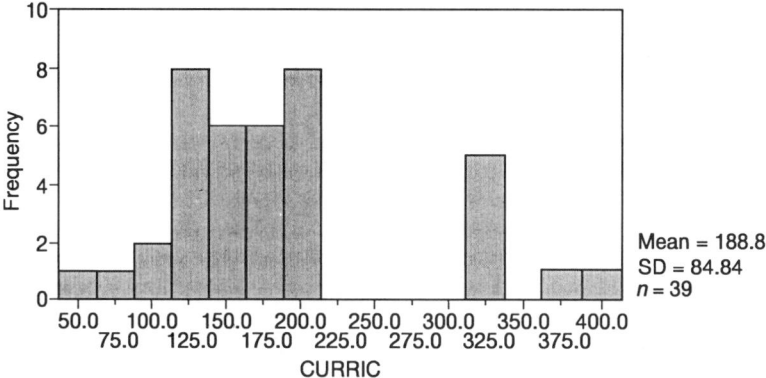

**Figure 2.2**  Histogram of curriculum coverage

Our second exploratory tactic is to look at the *scatterplot* of mathematics attainment against curriculum coverage. *Never do a regression without first doing a scatterplot.* Scatterplots, or scatter diagrams, can tell us about unusual points, known as outliers, and whether a straight line is an appropriate way of representing the relation between $y$ and $x$. At this stage, there is nothing in the scatterplot shown in Figure 2.3 that suggests that we should not fit a simple regression to these data. Thus, to start with, we fit a simple regression to the variables measured on their original scales. We can then look at the estimated residuals from the fitted model to see whether the model can be improved.

The fitted regression model, based on 39 pupils, is:

$$\text{MATH} = 13.1 + 0.083 \text{ CURRIC} \qquad (2.6)$$
$$(3.23) \quad (0.016)$$

The figures in brackets are the standard errors (s.e.) of the regression coefficients. The estimate of 13.1 for the intercept, $a$, tells us what we expect mathematics attainment to be when curriculum coverage is zero. More interestingly, the estimate of the slope, $b$, tells us that, for every unit change in CURRIC, MATH would be expected to change in the same direction by 0.083 units. Another way of putting this is to say that a standard deviation (SD) unit change in CURRIC predicts a change of 0.66 SD units in MATH. This is because, from Figures 2.1 and 2.2, we see that the SD of CURRIC is 7.9 times as great as the SD of MATH and so we multiply the estimate of $b$ by this ratio. This parameter (estimated to be 0.66 here) is sometimes called a *standardized regression coefficient* or a *beta coefficient*.

The estimate of $b$ is 5.3 times as great as its standard error, a ratio which is most unlikely to be consistent with a value of $b$ of zero in the population. In fact, the 95% confidence interval for $b$ is 0.051 to 0.11. However, the accuracy of this confidence interval does depend on the assumption that the residuals from the model are Normally distributed. We can also derive an approximate 95% confidence interval for beta, which is $0.051 \times 7.9$ to $0.11 \times 7.9$, or 0.40 to 0.91.

We also want to know how well the model fits the data. One way of assessing

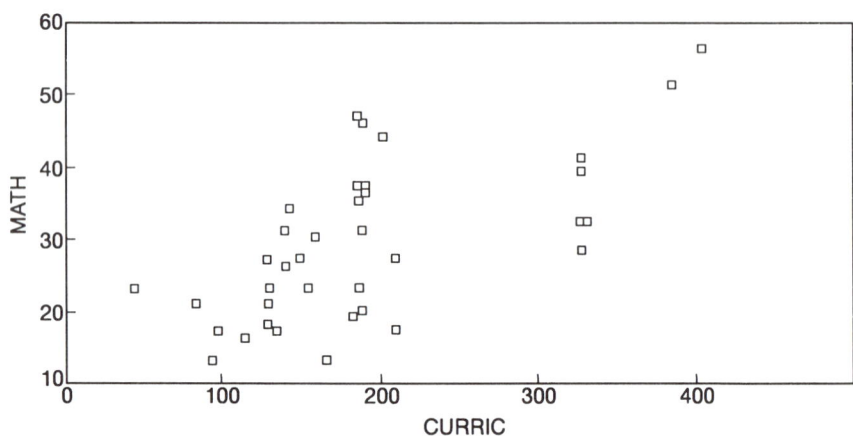

**Figure 2.3** Scatterplot of mathematics attainment by curriculum coverage

## 2.2 Simple regression with a continuous explanatory variable 15

this is to use the $R^2$ statistic; $R^2$ tells us how much of the variance in $y$ is explained by the regression (i.e. by $x$). For these data the estimate of $R^2$ is 0.43, so curriculum coverage explains 43% of the variance in mathematics attainment. For simple regression, $R^2$ is just the square of the correlation coefficient, $r$.

We now have the essential results from our simple regression analysis. However, we should not take our model on trust; instead, we should examine whether the model provides a satisfactory fit to the data, perhaps to improve it. The best statistical approach to model checking, or criticism, comes from examining the estimated residuals, $\hat{e}_i$, from the model where:

$$\hat{e}_i = y_i - \hat{a} - \hat{b} x_i \tag{2.7}$$

These are, in fact, estimated raw residuals. It is usually more helpful to look at estimated residuals which are standardized by their own standard errors. Hence, we will always use what are known as *studentized* residuals (Weisberg, 1980) in this chapter, which we label $\hat{r}_i$. In order for the estimated regression equation to have the desirable statistical properties mentioned earlier, we need an absence of pattern in our studentized residuals and an absence of any very large values of $\hat{r}_i$. in absolute terms.

The best way to look at the studentized residuals from a simple regression is to plot them against the predicted values of $y$ from the regression. The predicted values of the response we call $\hat{y}_i$, where:

$$\hat{y}_i = \hat{a} + \hat{b} x_i \tag{2.8}$$

and it is convenient to standardize $\hat{y}_i$ to have zero mean and unit variance.

This plot is shown in Figure 2.4. We see that the $\hat{r}_i$ are scattered around their mean of zero and none is very large. However, the variance of $\hat{r}_i$ shows some sign of increasing with the value of $\hat{y}_i$. An important requirement for a regression model to be a satisfactory one is known as *homoscedasticity* (or equal scatter). If the condition of homoscedasticity does not hold, so that instead the data are what is called heteroscedastic (or different scatter), and the variance of $\hat{r}_i$ is not constant but instead varies systematically with $\hat{y}_i$, then we need to consider a

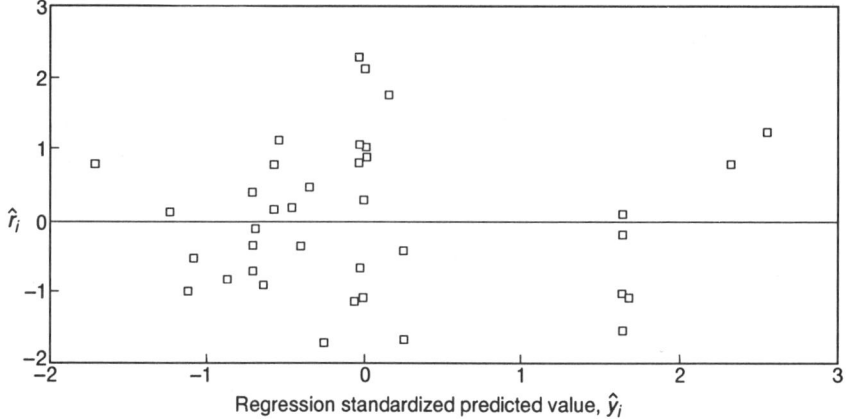

**Figure 2.4** Scatterplot of residuals, $\hat{r}_i$ by predicted response, $\hat{y}_i$

16  *Building the Foundations*

transformation of *y*. Let us now look in a little more detail at the issue of transforming variables when fitting statistical models.

## 2.3  Transforming variables

We saw in the first chapter (Section 1.3) that the scales we use to measure variables in educational research are often arbitrary. This can lead to problems of interpretation when thinking about the sizes of the effects of an explanatory variable on a response. However, if our scale is arbitrary then one order-preserving version of the scale is usually as good as another, and we can transform variables, in particular the response, to meet the assumptions of our statistical model, without sacrificing anything in terms of interpretation.

Thinking about the example in the previous section, we saw that there is evidence that the residuals are heteroscedastic, and that the distribution of the response is positively skewed. Often, a transformation which brings the distribution of the response closer to Normality also eliminates heteroscedasticity. In other words, Normalizing transformations are often variance-stabilizing transformations. One way of getting rid of heteroscedasticity is to use a log or square root transformation. When there is negative skew, squaring the values might help. Alternatively, we could just use a Normalizing transformation, replacing the ordered data values by the corresponding order statistics from a standard Normal distribution. One slight disadvantage of using a Normalizing transformation is that pupils getting the same score on the test do not get the same score on the transformed scale. For the data of the previous section, a log transformation works well, as we shall see.

If our response has a fixed and directly interpretable scale, then we may be willing to sacrifice some departure from Normality in order to retain ease of interpretation. For example, the amount of money spent by a local education authority is measured on a fixed scale which we will not always wish to lose by a non-linear transformation such as log or square root. In these cases, we have to find a balance between the assumptions of our model and the need to be able to give a clear interpretation of our findings from the model.

Often, it is convenient to work with a scale that has a mean of zero and a standard deviation of one. This is the case for the curriculum coverage variable of the previous section which we transform to ZCURRIC using the following linear transformation:

$$\text{ZCURRIC}_i = (\text{CURRIC}_i - \text{mean CURRIC}) / \text{SD (CURRIC)} \tag{2.9}$$

It is important to remember that these *z*-transformations are not Normalizing transformations (though, in many textbooks, Normalized transforms with mean zero and standard deviation one are denoted by the same symbol *z*).

Let us now see what happens when we take log mathematics attainment as our response and the *z*-transformed curriculum coverage as the explanatory variable. The fitted regression is now:

$$\text{LOGMATH} = 3.3 + 0.23 \text{ ZCURRIC} \tag{2.10}$$
$$(0.047) \quad (0.048)$$

The estimates of both *a* and *b* have, of course, changed because of the changes in

## 2.3 Transforming variables

scale. The estimate of $a$ tells us what the estimate of log mathematics attainment would be when the standardized value of curriculum coverage is zero, i.e. when the actual curriculum coverage has its mean value of 189 (see Figure 2.2). The standard deviation of LOGMATH is 0.37 so the ratio of the SD of ZCURRIC to the SD of LOGMATH is now $1/0.37 = 2.7$, and the estimated beta coefficient is therefore $0.23 \times 2.7 = 0.62$. The approximate 95% confidence interval for beta is 0.35 to 0.89. The estimate of $R^2$ is 39%.

These results are close to those obtained from the first model. However, we would now hope that the plot of the studentized residuals, $\hat{r}_i$, against the predicted values, $\hat{y}_i$, would have less pattern than before. Figure 2.5 suggests that this hope is realized. Also, to ensure the accuracy of any confidence intervals we estimate, we need the $\hat{r}_i$ to be Normally distributed. We can check this by plotting the ordered values of the $\hat{r}_i$ against the expected values if these ordered values came from a standard Normal distribution. This is known as a Q-Q (or *quantile-quantile*) plot. If the $\hat{r}_i$ are Normal then all the points would fall on the diagonal line. Figure 2.6 shows that they are very close to the diagonal, supporting the view that the $\hat{r}_i$ are Normal, and therefore making our inferences about the population coefficients more secure.

We see from Figure 2.5 that nearly all the values of $\hat{r}_i$ lie within two standard deviations of the mean of zero, as we would expect. We would be particularly suspicious of values of $\hat{r}_i$ greater than, say 3.5 or 4 (and less than $-3.5$ or $-4$) and would certainly want to investigate these data points in more detail, perhaps eventually deciding to omit them from the model. The absence of any pattern in Figure 2.5 is also consistent with the formulation of the model as a straight line. If the studentized residuals had, for example, shown the pattern in the hypothetical scatterplot of $\hat{r}_i$ against $x_i$ of Figure 2.7, then a quadratic model including $x^2$ as well as $x$ might have fitted better. Our original scatterplot (Figure 2.3) had already suggested that a model with non-linear terms in $x$ was unlikely to be needed.

There is one slightly strange feature of Figure 2.5, which was also seen in Figure 2.2. There is a gap in the plot on the horizontal axis with a cluster of

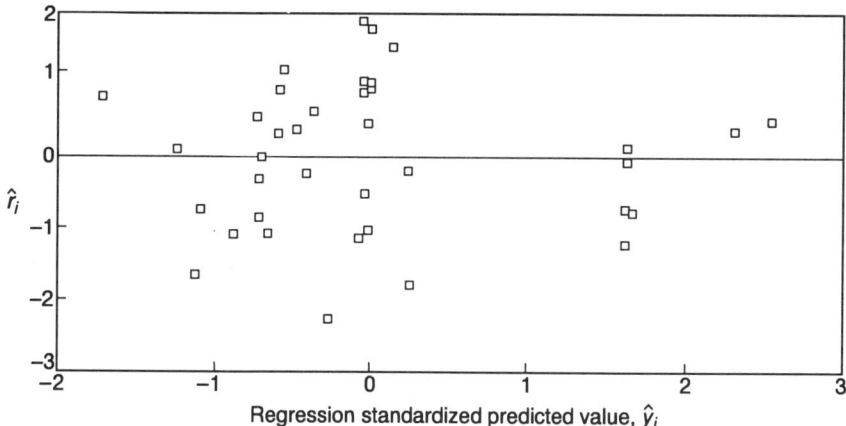

**Figure 2.5** Scatterplot of residuals, $\hat{r}_i$ by predicted response, $\hat{y}_i$

18  *Building the Foundations*

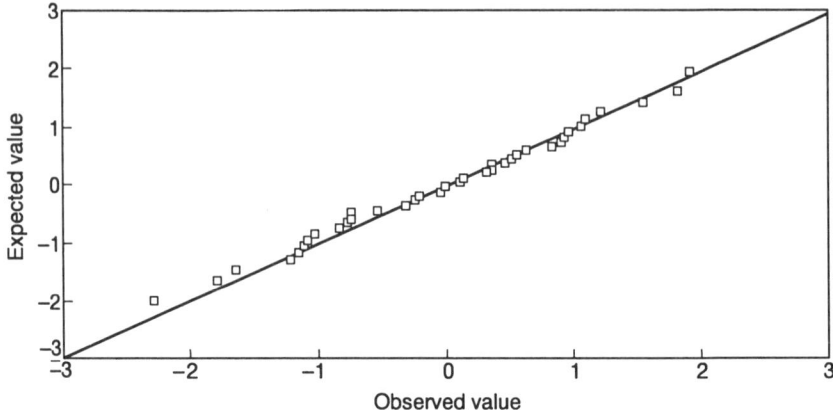

**Figure 2.6**  Q-Q plot of studentized residuals, $\hat{r}_i$

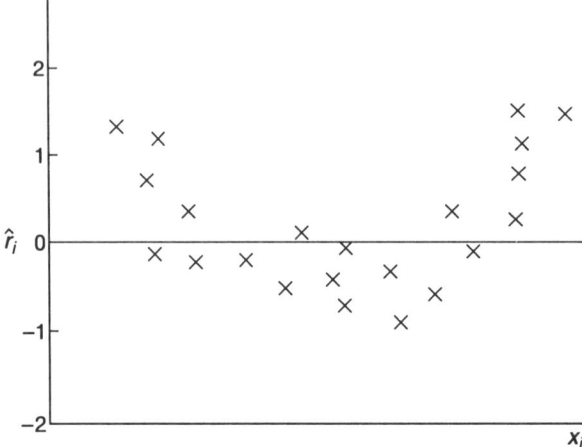

**Figure 2.7**  Studentized residuals, $\hat{r}_i$ by $x_i$

points with high values of $x$, somewhat removed from the rest of the data. We know that the data are a sample of pupils in three separate schools, rather than a simple random sample of pupils from a population. Perhaps these points are pupils in one of the schools, where curriculum coverage was higher. The grouping of pupils into classes and schools raises a number of very important issues for quantitative educational research, which will feature strongly in later chapters. For now, we consider how to investigate whether levels of mathematics attainment vary across the three schools in our sample.

## 2.4  Simple regression with a categorical explanatory variable

We represent unordered categorical explanatory variables in regression by using what are known as *dummy variables*. Suppose we want to relate mathematics

## 2.4 Simple regression with a categorical explanatory variable

attainment to sex, a binary explanatory variable. Then there is just one dummy variable, taking the value 0 for boys and 1 for girls (or the other way round, it does not matter). The regression method works in just the same way as described in the previous sections, and the interpretation of $b$ is that it gives the mean difference in mathematics attainment between girls ($x=1$) and boys ($x=0$). In other words, if $b$ is positive, then girls score higher than boys on average. Evidence that this is the case would be provided if $\hat{b} > 0$. Using regression in this way is like doing a $t$-test for the mean difference between two independent groups, but it is more informative and can be generalized easily to other, more complicated, situations.

Let us now extend the idea of dummy variables to a categorical explanatory variable with three categories, in our case school. We represent this classification by two dummy variables; the first dummy variable ($d_1$) takes the value 1 for school 1 and 0 otherwise, the second dummy variable ($d_2$) takes the value 1 for school 2 and 0 otherwise. This means we have coded school as follows.

| School | $d_1$ | $d_2$ |
| --- | --- | --- |
| 1 | 1 | 0 |
| 2 | 0 | 1 |
| 3 | 0 | 0 |

School 3 is therefore the baseline category, when both dummy variables are coded zero. This coding is, however, arbitrary and there is no particular reason for making school 3 the baseline category here. Dummy variables can be set up in the way that is most convenient, according to the aims of the study. For example, we might want the category with the most cases to be the baseline. However, the number of dummy variables must always be one less than the number of categories in order to be able to estimate the model (but see Exercise 2). It is wrong to code an unordered categorical explanatory variable with three categories as a single variable taking the values 0,1,2 (or 1,2,3 etc) as this implies an order which is not present in the variable.

The regression model relating mathematics attainment to school is:

$$y_i = a + b_1 d_{1i} + b_2 d_{2i} + e_i \qquad (2.11)$$

With the dummy variable coding used here, the estimate of $b_1$ gives the estimated mean difference in mathematics attainment between school 1 and school 3, whereas the estimate of $b_2$ gives the estimated mean difference between school 2 and school 3. The difference between $\hat{b}_1$ and $\hat{b}_2$ (i.e. $\hat{b}_1 - \hat{b}_2$) estimates the mean difference between schools 1 and 2.

Applying the above model to the data of the previous section and retaining the log transformation for the response, we end up with the fitted relationship:

$$\text{LOGMATH} = 3.25 + 0.18\,d_1 - 0.08\,d_2 \qquad (2.12)$$
$$(0.09) \quad (0.13) \quad (0.15)$$

In other words, mathematics attainment is estimated to be higher in school 1 (by 0.18 on the log scale) and lower in school 2 (by 0.08) than in school 3. The largest difference $(0.18 - (-0.08) = 0.26)$ is between schools 1 and 2. The estimated means for the three schools are 3.25 (school 3 – the baseline, represented by the intercept $a$), 3.43 (school 1) and 3.17 (school 2). However, the standard errors of the regression coefficients are quite high, suggesting that these differences may be no more than we would expect by chance, when there are no actual differences between the schools.

Rather than looking at the $t$-statistics for each dummy variable separately, we should assess the statistical significance of the model as a whole. We can do this by breaking down the total sum of squares, $\Sigma (y_i - \bar{y})^2$, into two components: the regression or explained sum of squares, $\Sigma (\hat{y}_i - \bar{y})^2$, and the residual, or unexplained, sum of squares, $\Sigma (y_i - \hat{y}_i)^2$. We then construct an *analysis of variance table* which, for these data, is:

|  | df | Sum of squares (SS) | Mean square (MS = SS/df) |
| --- | --- | --- | --- |
| Regression | 2 | 0.43 | 0.22 |
| Residual | 36 | 4.83 | 0.13 |
| Total | 38 | 5.26 |  |

There are two degrees of freedom (df) for the regression because there are two regression coefficients corresponding to the two dummy variables. The total df is always one less than the sample size because we estimate the mean of the response. The mean squares are calculated by dividing the sums of squares by df. The analysis of variance table generates two useful statistics. The ratio of the regression sum of squares to the total sum of squares is just $R^2$. For these data, $R^2$ is $0.43/5.26 = 0.08$, which is low (i.e. 92% of the variance of $y$ is not explained by this model). The ratio of the regression mean square to the residual mean square has an $F$ distribution (assuming the residuals are Normal) with the corresponding degrees of freedom. Here $F$ is $0.22/0.13 = 1.69$ with 2 and 36 df, which gives a $p$ value of about 0.20. Hence, there is no hard evidence from these data that mathematics attainment varies across these three particular schools.

An alternative approach to the analysis of the data discussed in this section is to use *one way analysis of variance*, rather than regression with dummy variables. The results would be identical, although analysis of variance (ANOVA) tends to emphasize the $F$ statistic rather than the regression coefficients. Regression does everything ANOVA does and more. Hence, I do no more than make passing references to ANOVA in this book.

We have seen how to estimate a simple regression model for data from three schools. If, however, we have 30 or 300 schools in our study, then a model with 29 or 299 dummy variables is, to say the least, cumbersome. In the next chapter, we will see how we can use multilevel models to deal with this kind of study. But for now we move on from simple to multiple regression.

## 2.5 Multiple regression with continuous explanatory variables

In Sections 2.2 and 2.3, we saw that our measure of curriculum coverage did account for some of the variation in mathematics attainment at the end of Year two, and that the size of the effect (around 0.6 SD units) was substantial. However, the finding is open to a number of different interpretations. One possibility is that teachers cover more of the curriculum with pupils who were doing well in mathematics at the *beginning* of Year two. Hence, we might find that once we take account of this initial attainment, then the link between attainment

## 2.5 Multiple regression with continuous explanatory variables

at the end of the year and curriculum coverage is less strong. To investigate this, we need a model with two explanatory variables; in other words, a multiple regression model with the following form:

$$y_i = a + b_1 x_{1i} + b_2 x_{2i} + e_i \qquad (2.13)$$

The response (LOGMATH) and ZCURRIC (previously $x$ and now $x_1$) are the same as before; $x_2$ is the measure of mathematics attainment at the beginning of the year, standardized for convenience (ZMATH1). The regression coefficients – $a$, $b_1$ and $b_2$ – are estimated as before using least squares, and the computation nowadays is always done using a statistical package such as SPSS, SAS and many others. All the results in this chapter come from using the regression procedures in SPSS for Windows.

We are particularly interested in the value of $b_1$ in this model because it gives us the size of the effect on end of year mathematics attainment of a unit change in curriculum coverage for *fixed* values of initial mathematics attainment. Putting this another way, it is the effect of ZCURRIC on LOGMATH net of the effect of ZMATH1 on LOGMATH. It is in this sense that we have taken account of, or controlled for (statistically rather than experimentally), the effect of initial mathematics attainment on later mathematics attainment. The value of $b_2$ tells us how much end of year mathematics attainment changes for a unit change in beginning of year mathematics attainment, holding curriculum coverage constant. At this stage, we assume that the effect of an explanatory variable does not vary with the values of the other explanatory variable. We relax this assumption in the next section.

The estimate of $b_1$ will be different from the estimate of $b$ in the simple regression equation in Section 2.3, because ZCURRIC and ZMATH1 are correlated, and also because $\hat{b}_2$ is not zero. Indeed, it is possible for $\hat{b}_1$ to be small even though $\hat{b}$ is not. Then all the apparent effect of curriculum coverage on mathematics attainment would, in fact, be taken up by the association between the two measures of mathematics attainment. A reasonable substantive conclusion in that case would be that teachers cover more of the curriculum with initially high attaining pupils, but this higher coverage does not result in these pupils being further ahead at the end of the year. In other words, there is no causal link between curriculum coverage and attainment.

Before we estimate the multiple regression model, we must plot LOGMATH against ZMATH1. The scatterplot is shown in Figure 2.8; it does not reveal anything untoward except that there is one pupil with a much higher value of ZMATH1 than any others. The inclusion of this possible outlier might unduly influence the estimates – a question we return to below.

We would like the correlation between the explanatory variables ZCURRIC and ZMATH1 not to be too high; if it were then we would be faced with a problem of collinearity between the explanatory variables, which leads to unstable estimates of $b_1$ and $b_2$. In fact, the correlation is 0.6 which is not so high as to cause problems. A correlation greater than 0.8 could be awkward and we would perhaps then have to drop one of the variables from the model.

Our fitted model is:

$$\text{LOGMATH} = 3.29 + 0.11 \, \text{ZCURRIC} + 0.20 \, \text{ZMATH1} \qquad (2.14)$$
$$\phantom{\text{LOGMATH} =\ } (0.04)\ (0.05) \phantom{\text{ZCURRIC} + } (0.05)$$

## 22 Building the Foundations

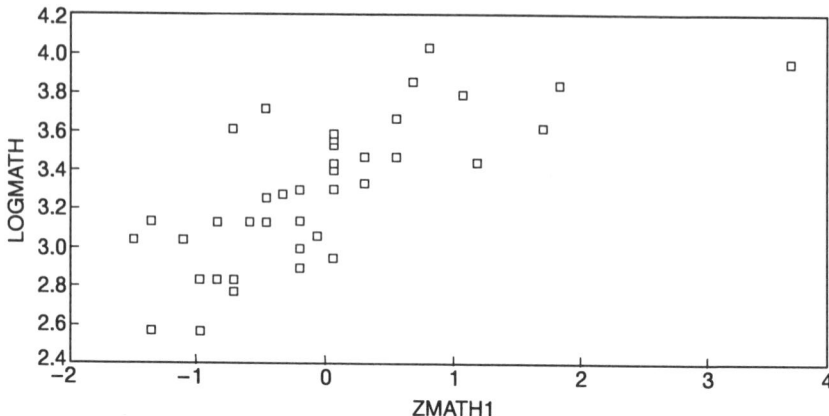

**Figure 2.8** Scatterplot of LOGMATH by ZMATH1

and the analysis of variance table is:

|  | df | Sum of squares | Mean square |
|---|---|---|---|
| Regression | 2 | 3.04 | 1.52 |
| Residual | 36 | 2.23 | 0.06 |
| Total | 38 | 5.26 |  |

giving an $R^2$ of 0.58 and an $F_{2,36}$ of 24.6, with a corresponding very small $p$ value. Clearly the model as a whole explains much more of the variation in LOGMATH than we would expect by chance.

We see that the introduction of ZMATH1 into our model has more than halved the coefficient of ZCURRIC, $b_1$ (from 0.23 to 0.11). We expect a change of only 0.30 (i.e. $0.11 \times 2.7$) SD units in LOGMATH for a standard deviation unit change in ZCURRIC, once we have controlled for ZMATH1. The 95% confidence interval for $b_1$ – from 0.0085 to 0.21 – is quite wide and almost includes zero. However, the combination of ZCURRIC and ZMATH1 has increased the value of $R^2$ from 39% to 58%.

Sometimes the value of $R^2$ is given too much emphasis in reports of multiple regression analyses, at the expense of the sizes of the coefficients themselves. It is the coefficients which tell us how our explanatory variables are related to the response. We cannot usually compare the magnitudes directly because the scales for the explanatory variables are, in general, different (although in this example they are not because they have been z-transformed). Instead we can use standardized, or beta, coefficients, as we have done here. These tell us how many standard deviation units change in $y$ we would expect for a standard deviation unit change in one of the explanatory variables, holding the other explanatory variables constant.

Returning to model checking, we should now plot $\hat{r}_i$ against each of the explanatory variables, as well as against $\hat{y}_i$, in order to be sure that we have an appropriate model. The residual plots (Figures 2.9–2.11) do not reveal any patterns, although they still suggest the presence of the same outlier found in

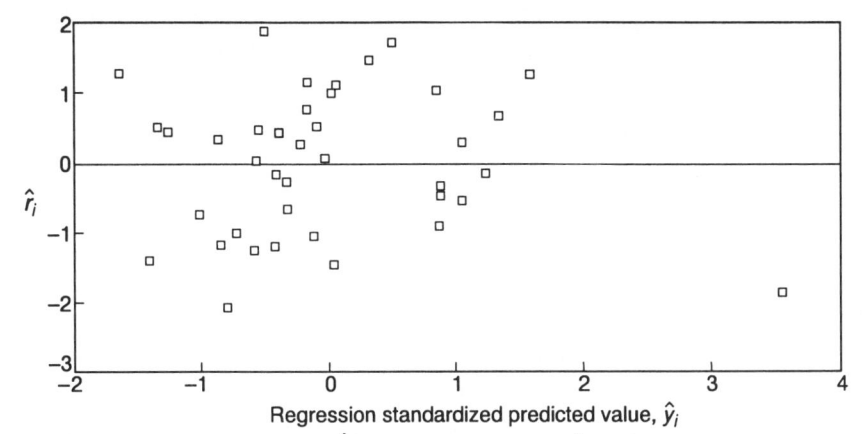

**Figure 2.9** Scatterplot of residuals, $\hat{r}_i$ by predicted response, $\hat{y}_i$

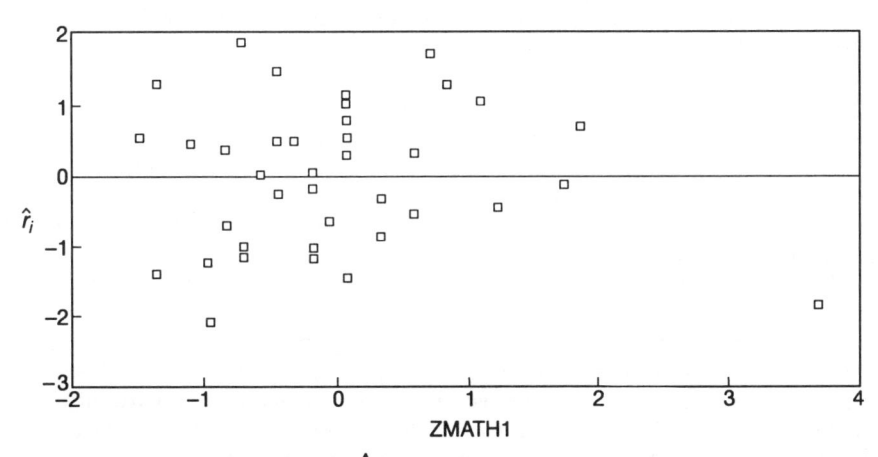

**Figure 2.10** Scatterplot of residuals, $\hat{r}_i$ by ZMATH1

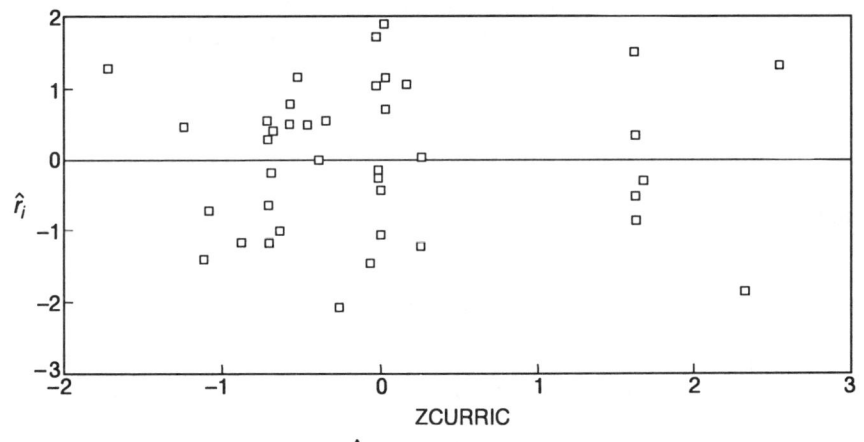

**Figure 2.11** Scatterplot of residuals, $\hat{r}_i$ by ZCURRIC

Figure 2.8. We also have another way of checking our model, and that is by looking at an influence statistic called Cook's $D$ (explained in Weisberg, 1980). In essence, Cook's $D$ ($D$ for distance) tells us how different our estimated regression coefficients would have been if each sample case were omitted in turn. The higher the value of $D$, the more likely it is that the case exerts considerable influence on the estimates of the coefficients. However, $D$ does not have a fixed range and so we focus on those values of $D$ which are considerably greater than, say, the 90th centile. From our sample of 39 pupils, only one has a value of $D$ much greater than the 90th centile, which is 0.08, and in that case (pupil 3), the outlier in our scatterplot has a value of $D$ of 0.75. If we omit that case from our analysis, we get the following results:

$$\text{LOGMATH} = 3.31 + 0.11 \text{ ZCURRIC} + 0.26 \text{ ZMATH1} \qquad (2.15)$$
$$(0.04)\ (0.05) \qquad\qquad (0.06)$$

These estimates – $\hat{b}_1$ and its s.e. (standard error) especially – are, in fact, very close to those obtained for the full sample, showing that although case 3 is different, it does not have a disproportionate influence on the results, and thus there is no reason to omit it from the model.

The model we have used in this section is, in fact, a model which is often used with longitudinal data. Here we have two measures of mathematics attainment, taken a school year apart on the same pupils. Essentially, we are interested in whether the level of curriculum coverage is related to pupils' *progress* over the school year. This is in contrast to the simple regression models of Sections 2.2 and 2.3, which were based on cross-sectional data and were only concerned with the relation with pupils' attainment at the end of the year.

The distinction between progress and attainment is an important one in educational research. Progress is a dynamic concept needing longitudinal data for its measurement, attainment is a static concept needing only cross-sectional data. It is usual to operationalize the concept of progress in the way we have here, looking at attainment at the second occasion *conditional* on attainment at the first occasion. An alternative approach is to measure progress in terms of a difference score over the period in question, but this becomes problematic when we do not have fixed scales over time. These issues come up again in Section 2.7, and in later chapters. A more detailed discussion of conditional and unconditional approaches to the measurement and analysis of change can be found in Plewis (1985).

## 2.6 Multiple regression with categorical explanatory variables

Suppose now that we have two categorical explanatory variables which, for simplicity, we suppose are binary. In particular, suppose we are interested in seeing whether we can explain the variation in mathematics attainment by the sex and ethnic group of the pupil. Our basic data are represented by a two by two table of means and standard deviations (Table 2.1).

The sample values in this table suggest that boys score higher than girls, white pupils higher than African Caribbean pupils, and that the sex difference is larger for white pupils than it is for African Caribbean pupils. We also see that the counts in the four cells are not equal, and are rather small. Also, there is little

## 2.6 Multiple regression with categorical explanatory variables

**Table 2.1** Log mathematics attainment (LOGMATH) by sex and ethnic group (means and SDs)

|  | Boys | Girls | Total |
|---|---|---|---|
| White | 3.50 (0.36) $n=17$ | 3.09 (0.28) $n=10$ | 3.35 (0.28) $n=27$ |
| African Caribbean | 3.23 (0.32) $n=6$ | 3.08 (0.34) $n=6$ | 3.16 (0.32) $n=12$ |
| Total | 3.43 (0.36) $n=23$ | 3.09 (0.29) $n=16$ | 3.29 (0.37) $n=39$ |

variation in the standard deviations across the cells. What the table does not tell us is how much of the variation in mathematics attainment is explained by sex and these two ethnic groups, and whether the observed differences across the cells are more than we would expect by chance. A multiple regression will tell us.

The model we use is, in fact, essentially the same as the one set out in Section 2.4, but with $d_1$ representing sex (0 = boys, 1 = girls) and $d_2$ representing ethnic group (EG; 0 = white, 1 = African Caribbean). The estimates are:

$$\text{LOGMATH} = 3.47 - 0.32\,\text{SEX} - 0.15\,\text{EG} \qquad (2.16)$$
$$(0.08)\ (0.11) \qquad (0.12)$$

with $R^2$ equal to 0.24 and $F_{2,36} = 5.78$, $p < 0.001$. The estimate for sex (−0.32) gives us the estimated mean difference between boys and girls, having allowed for the ethnic group difference. It corresponds to an *effect size* of −0.32/0.37 = −0.86, where the effect size here is the difference between boys and girls in SD units of LOGMATH. The estimate for ethnic group (−0.15) gives us the estimated mean difference between the two groups having allowed for the sex difference, and corresponds to an effect size of −0.41. (Effect size is *not* the same as the beta coefficient; the estimated beta coefficients here are −0.43 and −0.19.) In combination, our two binary explanatory variables account for nearly a quarter of the variation in LOGMATH, and the regression as a whole is statistically significant. However, it does appear that sex is much the more important of the two explanatory variables.

The previous model is a *main effects* model; it does not allow for the possibility that the sex difference is not the same for the two ethnic groups (or, equivalently, that the ethnic group difference is not the same for boys and girls). A more complete model would incorporate an interaction between sex and ethnic group. It is straightforward to include interactions in multiple regression models. Quite simply, two-way, or first-order, interaction terms are just the product of the two explanatory variables in question. In this case, we create a new variable, $d_3$, which equals $d_1 d_2$, and which therefore takes the value one for African Caribbean girls and zero for the other three groups. Our fitted model is now:

$$\text{LOGMATH} = 3.50 - 0.41\,\text{SEX} - 0.27\,\text{EG} + 0.26\,\text{SEX.EG} \qquad (2.17)$$
$$(0.08)\ (0.13) \qquad (0.16) \qquad (0.23)$$

and the $F_{3,35}$ statistic is now 4.27, $p < 0.02$. (Note the change in the degrees of freedom for $F$.)

Models which include interactions are more difficult to interpret than main effects models. No longer can we discuss sex differences on their own; instead, we must describe them within each ethnic group. Consequently, here the difference between white boys and girls is –0.41, or 0.41 in favour of the boys when SEX.EG is zero (i.e. for white pupils). However, for African Caribbean pupils, the sex difference is only 0.15 (–0.41 + 0.26) when SEX.EG is one (as it is for African Caribbean girls). With these data, however, the standard error for SEX.EG (0.23) is large relative to the coefficient (0.26), and so we would be justified in concluding that a main effects model is sufficient for these data. One important point to remember is that we cannot interpret a multiple regression model which includes an interaction but which does not include the corresponding main effects.

Suppose our categorical explanatory variables are not both binary. For example, we might want to construct a model relating mathematics attainment to sex, school and the interaction between them. Our underlying question might be to discover whether the sex difference found in the previous model is the same magnitude for each of the three schools. This is a reasonable question to ask even though there are no overall differences between the schools. The model is then:

$$y_i = a + b_1 d_{1i} + b_2 d_{2i} + b_3 \text{ SEX} + b_4 d_{1i} \cdot \text{SEX} + b_5 d_{2i} \cdot \text{SEX} + e_i \qquad (2.18)$$

This looks rather complicated because, as well as having two coefficients for the main effect of school ($b_1$ and $b_2$) and one for the main effect of sex ($b_3$), we must also have two terms representing the interaction between sex and school. The number of degrees of freedom, and hence the number of terms for an interaction, is equal to the product of the degrees of freedom for each main effect included in the interaction. In fact, when we estimate this model, *all* the coefficients are small compared with their standard errors, partly because the interaction is unimportant and partly because we are estimating rather a lot of coefficients (five in all) for a rather small sample of 39.

In principle, we could extend the model further to include ethnic group. A full model would then have 11 terms in total: three main effects (four terms), three two-way interactions (five terms altogether), and one three-way interaction (sex by ethnic group by school, represented by two terms). Just as we must not include interactions without the main effects, so we must not include three-way interactions without the subset of two-way interactions. However, three-way interactions can sometimes be accounted for by one or two influential points, and usually we will prefer a model which is more parsimonious than one including a lot of two and higher-way interactions.

## 2.7 Multiple regression with a mixture of categorical and continuous explanatory variables

Let us now combine the models of the previous sections, to consider situations with at least one continuous and at least one categorical explanatory variable. As we shall see, the simplest version of this model is an especially important one for the analysis of data from observational studies, including quasi-experiments. For now, suppose we are interested in relating end of the year mathematics attainment to earlier mathematics attainment and sex. We have already found that ZMATH1 is a good predictor of LOGMATH, and that boys are ahead of girls at the end of the year. What we would now like to see is whether boys make

## 2.7 Multiple regression with a mixture of explanatory variables

more progress than girls during the year. In other words, is the sex difference in attainment at the end of the year wholly attributable to a difference in attainment at the beginning of the year. Our fitted model turns out to be:

$$\text{LOGMATH} = 3.38 + 0.24 \text{ ZMATH1} - 0.21 \text{ SEX} \qquad (2.19)$$
$$(0.05) \ (0.04) \qquad\qquad (0.08)$$

and shows that, indeed, boys do make more progress in mathematics during the year. The effect of sex on progress is smaller than its effect on attainment – the estimated effect size falls from 0.86 to $0.21/0.37 = 0.57$ SD units – but it is still more than we would expect by chance, as the $t$ ratio for SEX is $-2.55$.

This model is often known as the *analysis of covariance* (ANCOVA) model, especially when analysing experimental data. It does, however, assume that the relation between the two measures of mathematics attainment is the same for the two sexes; in other words, there is no interaction between ZMATH1 and SEX. However, we might have a situation like Figure 2.12. This would imply that, rather than there being a constant difference between boys and girls, the difference varies according to the value of the first measure of attainment. In Figure 2.12, the difference actually changes sign: boys make more progress for low values of initial attainment, girls make more for high values. But, for the data of the example, the estimated coefficient for the interaction between ZMATH1 and SEX is small – it equals 0.044 with a standard error of 0.093. Hence, we can reasonably conclude that boys make the same amount more progress than girls over the whole range of the initial attainment measure, as far as this small dataset goes.

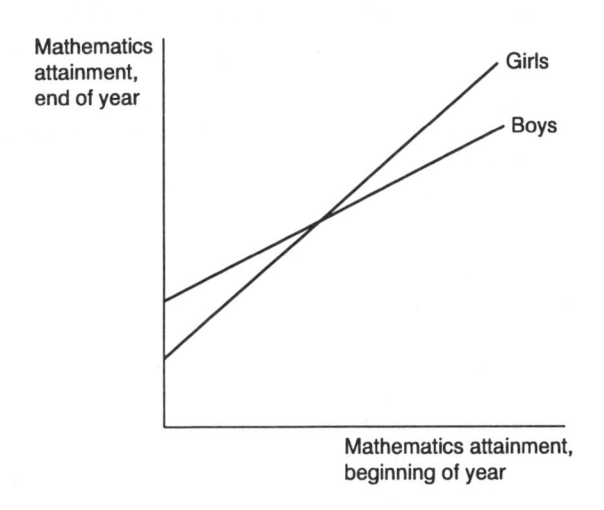

**Figure 2.12** Mathematics progress interacting with sex

If our two measures of attainment were pre and post measures, and if, rather than sex, our binary variable was a variable representing two groups – an intervention and a control group – then we have an appropriate model for the analysis of a common quasi-experimental design. The model is:

$$y_i = a + b_1 x_{1i} + b_2 d_i + e_i \qquad (2.20)$$

where $y$ is the post-test, $x_1$ is the pre-test and $d$ is a dummy variable (0=control, 1=intervention). The value of $b_2$ is a measure of relative change which, if all the differences between the two groups are taken up by the pre-test, can be interpreted causally as an intervention effect. Usually, however, when there is no randomization, the two groups differ on several variables, and not just on the pre-test. All these so-called confounding variables need to be included as explanatory variables. We take up further discussion of this controversial area of applied statistics in Chapters 3 and 4.

## 2.8 The assumptions of regression

As we now know, the regression model, like all statistical models, is based on a series of assumptions about the data to which it is applied. These assumptions have been brought out in the context of a series of examples in previous sections of this chapter. They are re-examined here in a slightly more formal way than hitherto.

The main assumptions concern the nature of the residuals, $e_i$. First, we assume that their variance is constant, i.e. that they are homoscedastic. Another way of saying this is that the variance of the response, conditional on the explanatory variables, should be constant. If we assume the residuals are homoscedastic when, in fact, they are not, then we will underestimate the standard errors, or overstate the precision, of our estimated regression coefficients. This, in turn, will mean that our confidence intervals will be too narrow. Some evidence for this was found in Sections 2.2 and 2.3; the confidence interval for the beta coefficient for curriculum coverage was a little wider after transforming the response with a log transformation to reduce heteroscedasticity.

The second assumption we make about the residuals is that they are independent of each other. This is guaranteed if our sample is a simple random sample from the population of interest. However, if our observations are clustered, for example by school, then the assumption of independence is no longer tenable. Nor is it if the observations are a time series of some kind – the residuals are then autocorrelated. Assuming independence when it does not hold will again lead to estimates whose precision is overestimated. We return to this issue in the next chapter.

A necessary condition for a regression coefficient to have a causal interpretation is that the residuals are independent of the explanatory variables in the model. This is not an easy condition to satisfy in educational research that is not experimental. It is, however, especially important when we are analysing intervention studies of the kind described in the previous section. We can give a causal interpretation to the coefficient attached to the binary intervention variable if and only if we have specified the model sufficiently well that any omitted variables are independent of the explanatory variables in the model (or if we have used randomization in the design).

Finally, the residuals should be Normally distributed in order for any statistical inferences to be firmly grounded. However, some, not too severe, departures from Normality are unlikely to have any important effects because the $t$ and $F$ distributions used to make inferences from regression models are relatively robust. On the other hand, especially discrepant observations can have an undesirable influence on the fitted model.

The other kind of assumptions in regression are those relating to the way the response is linked to the explanatory variables. Is it reasonable to suppose that the relation between $y$ and $x$ is a straight line or would a quadratic ($x^2$) or an exponential function ($\exp(x)$) give a better fit? Similarly, do the effects of the explanatory variables combine in an additive way or are interactions needed?

The regression model is quite flexible in its ability to handle different functions. For example, a model such as:

$$y = a \exp(b_1 x_1) \exp(b_2 x_2) \exp(u) \tag{2.21}$$

can, by taking logs to base e of both sides, be converted to:

$$\log_e y = \log_e a + b_1 x_1 + b_2 x_2 + u \tag{2.22}$$

where $u$ is the residual. Because $\log_e a$ is a constant, this is just a standard multiple regression model with a log transformation of the response.

On the other hand, models such as:

$$y = a \exp(b_1 x_1) + b_2 x_2 + u \tag{2.23}$$

are non-linear models which cannot be transformed to a form we recognize, and which therefore require special methods of estimation beyond the scope of this book.

As we have seen, we can check the validity of most of our assumptions with scatterplots, and with different kinds of residual plots such as Figures 2.9–2.11.

## 2.9 Other topics in regression

The previous sections have covered the most important aspects of statistical modelling using regression. One issue which has not been covered is measurement error – the fact that often in educational research, explanatory variables are unreliable. This important, but often ignored, topic is discussed in Chapter 8. However, there are a number of other topics which are considered briefly here.

### 2.9.1 Automatic selection of explanatory variables

Sometimes we have a large number of potential explanatory variables that could be fitted to the data. This can happen in a big study, when there is little theoretical background to guide the choice of explanatory variables. One popular, albeit rather dangerous, way of dealing with this situation is to let the computer choose the most 'important' explanatory variables using what are known as *step-wise* procedures. Here, importance is defined solely in terms of variability explained – variables are added to the model in order of how much variance they explain, as long as they explain more variability in $y$ than would be expected by chance. The disadvantages of this approach are that the chosen model can vary according to whether step-wise forward or step-wise backward methods are used, that variables which might be important theoretically will not necessarily appear in the final model, and that interactions are ignored. An alternative to step-wise procedures when there is a large pool of explanatory variables is to reduce this pool to a few linear combinations, these linear combinations being the most important *principal components* of the explanatory variables. This method is described in, for example, Weisberg (1980). It has the disadvantage that the principal components may not easily be interpretable; the method is probably more useful when the aim of the regression model is to predict the response rather than to explain it.

### 2.9.2 Prediction

We saw in Section 2.2 how predicted values of the response are used in model checking. We might also want to predict the value of the response for values of explanatory variables which were not observed in the sample. The method is the same as we used in Section 2.2 – we substitute the new values of the explanatory variables to get a predicted value of $y$. However, we need to bear in mind a number of points before we conclude that we have a useful prediction. First, if the value of $R^2$ is small for the model as fitted to the sample data, we clearly cannot place much faith in any predictions, because there are other important but omitted explanatory variables which affect the response. Secondly, if we have a large random sample from the population, our predictions for that population may be reasonable, but predictions for other populations will not necessarily be so. Thirdly, and very importantly, if we do not have a large random sample from the population, then we must be very wary about making predictions which use values of any explanatory variables falling outside the range observed in our sample. This is known as the problem of extrapolation. Finally, any predictions we make will be subject to prediction error. The further the value of the explanatory variable from its mean, the greater the prediction error.

### 2.9.3 Missing data

It is usual to have missing data in quantitative educational research, and it is common to ignore the problem, and just to analyse the cases with complete data. This is often known as *listwise deletion*. If the total amount of missing data is small, and if the data that are missing are a random subset of what would have been observed if there had been no missing cases, then complete case analysis will be reasonable. Unfortunately, it is more usual for neither of these two conditions to hold. This is especially so with longitudinal data. Most of the research on methods for dealing with missing data in regression has focused on missing values of the explanatory variables. A number of ways of basing analyses on all the cases, rather than just on the complete cases, have been proposed, some of which are statistically sophisticated. Little and Schenker (1995) give a review. The particular problems posed by missing data in longitudinal studies are discussed in Chapter 4.

### 2.9.4 Ecological fallacy

Suppose, instead of having data on individual pupils on mathematics attainment and curriculum coverage, as in the earlier sections, we only have data at the classroom level. That is, we only know the mean mathematics attainment and mean coverage for each classroom. We could still fit a regression line to the classroom data, but any relation we find at the classroom level does not allow us to draw conclusions about what is happening at the pupil level. In other words, we must be wary about making inferences about individual behaviour and individual processes from aggregate data. If we do, we run the danger of committing the *ecological fallacy* as discussed by Freedman *et al.* (1991). Figure 2.13 illustrates the problem. *Within* each of the three groups, there is a slight *negative* relation between $y$ and $x$; as pupils' curriculum coverage goes up, their mathematics attainment goes down. However, *between* the three groups there is a marked *positive* relation; as classroom mean curriculum coverage goes up, so does mean mathematics attainment. In other words, there appear to be two different processes, both of which we want to be able, ideally, to model in our analyses.

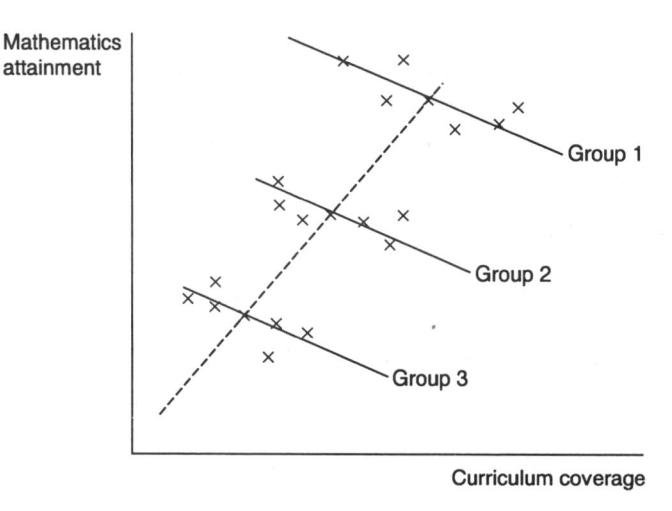

**Figure 2.13** Illustrating the ecological fallacy: different between group (- - -) and within group (——) regression

This is a good point at which to end this chapter, because the next chapter is explicitly concerned with methods that take into account the structured or hierarchical nature of educational data.

## Exercises

Answers to those exercises marked '#' can be found at the end of the book.

**1** Using dataset 2.1 on the disk, fit the model of Section 2.3 (p. 16) but using a square root transformation of mathematics attainment at the end of Year two (MATH) rather than a log transformation. Compare the two models in terms of beta coefficients, residual plots and $R^2$, and comment on your results.

Under what circumstances would a square root transformation be preferable to a log transformation?

**2** From the model in Section 2.4 (p. 19) deduce what the estimated regression coefficients would be:
  (a) if school 1 were used as the baseline category, and
  (b) if the constant, $a$, were omitted from the model and three dummy variables were used instead.
Use dataset 2.1 on the disk to obtain the standard errors for each of the estimated coefficients for the two models.
(#)

**3** Draw a sketch showing an interaction effect between two binary explanatory variables on a continuous response when the two main effects are both zero.
(#)

**4** Write down and fit a model to dataset 2.1 with log of mathematics attainment at the end of Year two as the response, and $z$-transformed mathematics attainment at the end of Year one and school as the explanatory variables. Comment on your results, and compare them with the answer to Exercise 2.

**5** The following table gives the means and standard deviations of class size and reading attainment for Year two pupils in four different countries.

| Country | Class size | Reading attainment |
|---|---|---|
| A | 24 (8) | 100 (15) |
| B | 27 (6) | 110 (18) |
| C | 32 (9) | 115 (22) |
| D | 40 (4) | 118 (30) |

What conclusions can you draw from these data? What further data might you need to be able to learn more about the relation between class size and reading progress?

# 3
# Populations with Structure: Multilevel Models for Continuous Responses

## 3.1 Introduction

There were indications in the previous chapter that the models described there would not always be appropriate for the kinds of data commonly found in educational research. We saw, in Section 2.4, how to relate attainment to a set of three schools using dummy variables. This works well when there are only a few schools in the sample but does not conveniently extend to studies with a reasonable number of schools. In Section 2.9.4, we noted that within group regressions can differ from between group regressions, and that these differences could correspond to different, and complementary, educational processes. Hence, we often need a rather more sophisticated approach to statistical modelling when we analyse educational data. This chapter serves as an introduction to this approach, known as multilevel modelling, while at the same time building on the ideas presented in Chapter 2.

This chapter starts with a description of what we mean by a hierarchical data set, and why it is important to take account of hierarchical structures in educational research. We then move through a series of multilevel models, starting with very simple models for measuring between school and between teacher differences, moving on to simple models which can be used in studies of school and teacher effectiveness, and finally considering models for measuring the effects of educational contexts and interventions. The emphasis throughout is on the way we specify models, and on the way we interpret the results from them, rather than on statistical theories of estimation, and computational methods.

## 3.2 The hierarchical nature of educational data

It is useful to think of modern educational systems as having several layers, these layers defining a pyramid with central government as the apex and pupils forming the base (Figure 3.1). The pyramid in turn defines a hierarchy; each school belongs to one and only one local education authority (LEA), each teacher belongs to one and only one school, and each pupil belongs to one and only one classroom (or subject teacher). We call this a *hierarchical* or *nested* structure. At each level of the pyramid below the apex, there is variability in most things we measure – between LEAs, between schools within LEAs and so on. The existence of these variabilities in the population has very important implications for the way in which we design research studies, and for the way in which we interpret results from them. Moreover, the extent to which we represent, or sample from, each of these levels in our design should influence the way in which we analyse our data.

For many years, the structure of educational data was ignored in large-scale quantitative educational research. Instead, the data were treated as if they came from a population with no structure. This happened partly because researchers, and the authors of the statistical textbooks they consulted, were unaware of the implications of so doing, partly because the statistical models they needed had not been developed. Matters came to a head in 1976, with the publication of a controversial study on teaching styles and pupil progress (Bennett, 1976). This study purported to show that pupils made more progress in primary school if they were taught by a 'formal' rather than by an 'informal' or 'progressive' teacher. Leaving aside the very real problems of creating a classification of teaching styles, the main problem with Bennett's analysis is that he ignored variability between teachers in pupils' progress. This, in turn, led him to overstate the precision of the observed differences between teaching styles. This was demonstrated in a re-analysis of the Bennett data by Aitkin *et al.* (1981), which was one of the first examples of multilevel modelling in practice, and which gave a real impetus to the development of these models in education, and in many other fields too. An overview of the controversies surrounding the Bennett study, and two other studies published in the late 1970s, can be found in a pamphlet by the Radical Statistics Education Group (1982).

Multilevel modelling of continuous responses is an extension of the regression

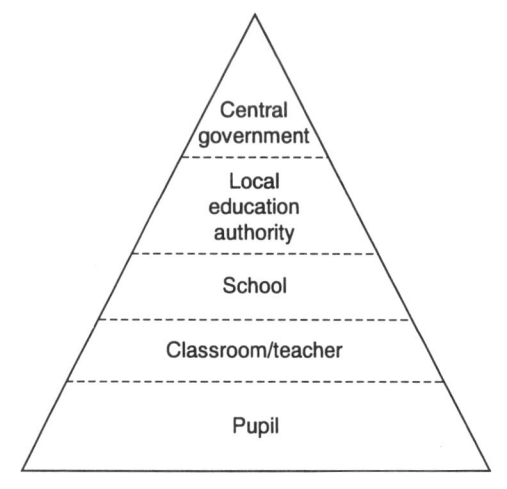

**Figure 3.1** The hierarchical structure of educational data

models of the kind described in Chapter 2, and has two particular strengths. First, by taking account of *all* the variability in the data, the standard errors of the regression coefficients in multilevel models are correctly estimated. The precision of estimated coefficients from single level regressions tends to be overstated, i.e. the standard errors are underestimated. This was the thrust of the Aitkin *et al.* (1981) paper. Secondly, and perhaps more importantly, a multilevel modelling approach enables researchers to ask, and to answer, questions about educational processes which were ignored in the past. They can, for example, ask whether the relation between attainment and curriculum coverage varies from teacher to teacher. Moreover, they can also ask *why* this might be so. In that sense, multilevel modelling offers researchers the possibility of being much more imaginative theoretically, as well as when they analyse their data. It also enables researchers to avoid asking the question – 'at what level should I analyse my data?' – and so to avoid the problems and pitfalls of aggregation, and the ecological fallacy of Section 2.9.4. The point about multilevel modelling is that data from all available levels can, and should, be analysed. The following sections of this chapter, and all the next chapter, illustrate these advantages. However, there will always be some datasets, especially small datasets with very few higher level units such as schools, for which the simpler approaches of Chapter 2 will be appropriate.

## 3.3 Partitioning variability by level

In Section 2.4, we saw how to look at mean attainment differences between three schools using dummy variables in a simple regression model. We also recognized that the approach did not easily extend to datasets with large numbers of schools. Statisticians often refer to the estimates attached to each school from the dummy variable regression as *fixed effects*. But with many schools, especially if these schools are randomly selected from a population, it is more convenient to think about 'school' as a *random effect*. (School is used here as shorthand for any kind of institutional grouping.) Indeed, multilevel models are sometimes described as random effects models. One implication of the move from fixed to random effects models is that rather than describe differences between individual schools, we think in terms of the variance of the response between schools. This, in turn, implies that rather than having to estimate a number, sometimes a substantial number, of fixed parameters (one less than the number of schools), we estimate just one variance. The fewer parameters we have to estimate from a dataset the better, other things being equal.

Let us illustrate these ideas with a simple example (dataset 3.1 on the disk). We have a sample of 777 pupils from 20 inner London primary schools, whose reading attainment was assessed at the end of Year one. The data come from a longitudinal study of school differences described in Plewis (1991a). For now, all we want to know is how much of the total variance in reading attainment can be attributed:

(i) to variation between schools, and
(ii) to variation between pupils within schools. (There is only one LEA in the study.)

We are also interested in whether the between school variance is more than we would expect by chance. We have just two levels: school and pupil. Hence, our model looks like this:

$$y_{ij} = b_0 + u_j + e_{ij} \qquad (3.1)$$

(It is convenient to use $b_0$ for the constant rather than $a$ as we did in Chapter 2.)

One important aspect of this model, which distinguishes it from those in Chapter 2, is that there are now two subscripts; $i$ for pupil (level one) and $j$ for school (level two). There are 20 schools so $j$ goes from 1 to 20. The number of pupils in each school varies, from a low of 22 to a high of 51. On the left-hand side of the model, $y_{ij}$ is reading attainment for pupil $i$ in school $j$. On the right-hand side, $b_0$ is just the overall mean and $u_j$ is the term for the effect or departure of the $j$th school from the overall mean. We treat $u_j$ as a random variable and so we want to estimate its variance ($\sigma_u^2$), assuming its mean is zero. Also, $e_{ij}$ is the residual term for pupils with a mean of zero and a variance $\sigma_e^2$. It is customary to assume that the $u_j$ have a Normal distribution, and we continue to assume, as we did in Chapter 2, that the $e_{ij}$ are Normal.

There are three parameters to estimate:

(i) the overall mean, $b_0$,
(ii) the between school variance in reading attainment, $\sigma_u^2$, and
(iii) the between pupil within school variance, $\sigma_e^2$.

There are many ways to estimate these parameters but we use the method of iterative generalized least squares incorporated into the package MLn (Rasbash and Woodhouse, 1995). The technical details of this method need not concern us here; they can be found in Goldstein (1995). Further discussion of statistical packages for multilevel modelling can be found at the end of this chapter.

Applying MLn to this simplest of two-level models, our results are as follows, with standard errors in brackets:

(i) $\hat{b}_0 = 3.39$ (0.10) so the overall mean for reading attainment is 3.39 on this scale (which is, in fact, the square root of the original scale, the distribution of which was skewed);
(ii) $\hat{\sigma}_u^2 = 0.16$ (0.06);
(iii) $\hat{\sigma}_e^2 = 1.4$ (0.07);

so the between school variance is about 10% of the total (0.16/(1.4 + 0.16)). The ratio of level two variance to total variance is often called the *intra class*, or *intra unit correlation*, $\rho$ (rho). Although the usual approximation of comparing an estimate with twice its standard error does not work well with variances, the between school variance here is more than we would expect by chance. In other words, the reading attainments of pupils in the same school are more alike on average than the attainments of pupils in different schools. It is rare for this not to be so, and this has important implications for those statistical analyses which assume simple random sampling and so ignore the lack of independence between pupils. On the other hand, an estimate of $\rho$ of 0.10, as here, is on the low side. The intra unit correlation would be one if all the pupils in a school had the same score, and zero if there were no differences in mean attainment between schools.

## 3.3 Partitioning variability by level

The description of the dataset given above is incomplete; the pupils were in fact grouped into 46 classrooms, with most schools having two classrooms and a few having three. Hence, our previous estimate of between pupil within school variance will include some between classroom within school variance as well. If we repeat the previous analysis with classroom, rather than school, as level two, we get:

$$\hat{\sigma}_u^2 = 0.25\ (0.07);\ \hat{\sigma}_e^2 = 1.3\ (0.07).$$

In other words, as expected, the between classroom variance is greater than the between school variance, and the between pupil within classroom variance is less than the between pupil within school variance. (The total variance is, of course, unchanged.) Now, the level two variance is 16% of the total ($\hat{\rho} = 0.16$), implying that the reading attainment of pupils is more alike within classrooms than it is within schools.

We are working with a very simple model here. Nevertheless, we should still subject it to the kinds of checking we met in Chapter 2. In particular, we should look carefully at the estimated level two residuals to see if there are any classrooms which are outliers, and to check whether the distribution of these residuals is approximately Normal. These residuals are easily generated by MLn and Figure 3.2 gives the histogram of the standardized residuals. They appear to be well behaved.

There is, however, one rather important property which level two residuals have, which makes them different from the estimates which would have been obtained from a fixed effects (or dummy variable) regression model. They are what is known as *shrunken* estimates, which means that they are generally closer to zero than the fixed effects estimates. The standard deviation of these shrunken residuals is 0.43, whereas the standard deviation of the classroom means is 0.64. The shrinkage is generally greater for classrooms with few pupils. For example, one classroom in the study with only two pupils in our sample has a mean 0.58 units below the sample mean but a level two residual of only −0.17. Essentially, the logic of shrinkage is that when estimates of level two residuals are based on few observations then they are subject to the vagaries of sampling

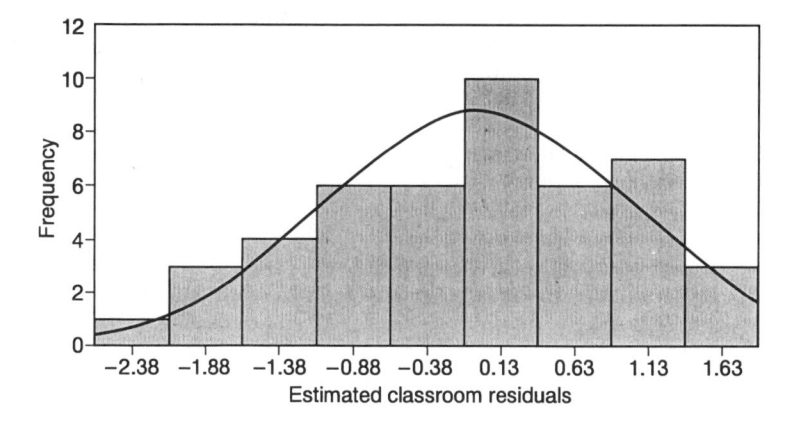

**Figure 3.2** Histogram of standardized classroom residuals

more than if they are based on many observations. Therefore, it makes sense for these level two units to 'borrow' information from all the other level two units to mitigate this sampling variability. The borrowed information pulls the estimated residual towards the mean of zero.

So far, we have fitted two models, each of which has just two levels. In the first model, the between classroom variance was subsumed in the between pupil variance because classroom was not included as a level. In the second model, the between school variance was subsumed in the between classroom variance because school was not included as a level. We really want a three-level model here – school, classroom and pupil – to represent properly the structure of the data. We write this model as:

$$y_{ijk} = b_0 + v_k + u_{jk} + e_{ijk} \qquad (3.2)$$

with the third subscript, $k$, representing the third level, school, and with three random effects – $v$ for school, $u$ for classroom within school and $e$ for pupil within classroom – to represent the three sources of variation. Here, $y_{ijk}$ is the reading attainment of pupil $i$ in classroom $j$ in school $k$.

Fitting three-level models is very easy, using just the same principles as for two-level models. We find, for these data, that the between classroom variance of the two-level model divides roughly equally into between school variance ($\hat{\sigma}_v^2 = 0.12$, s.e. $=0.07$) and between classroom within school variance ($\hat{\sigma}_u^2 = 0.14$, s.e. $=0.06$). The estimated between pupil variance is unchanged at 1.3 (s.e. $=0.07$). We should note, however, that the confidence intervals for the two variance components are wide, and so the data are consistent with a markedly different partition of between school and between classroom variance from the one observed here.

We now move on from these simple multilevel models, which merely partition the total response variance into different components at different levels, to models with at least one continuous explanatory variable measured at level one.

## 3.4 Multilevel models for school and teacher effectiveness

The previous section focused on variation in attainment at the end of Year one by school and classroom, as well as by pupil. Now we introduce a second measure of pupil reading attainment, collected a year earlier at the end of the reception year, as an explanatory variable. In other words, we focus on progress during Year one. (The distinction between attainment and progress was explained in the previous chapter on p. 24.) We want to know whether there are any differences between schools, and between classrooms, in the amount of reading progress made by pupils. Let us follow the same pattern of the previous section, and look first at between school differences. Our model is:

$$y_{ij} = b_0 + b_1 x_{ij} + u_j + e_{ij} \qquad (3.3)$$

or, equivalently,

$$y_{ij} = b_{0j} + b_1 x_{ij} + e_{ij} \quad \text{with} \quad b_{0j} = b_0 + u_j \qquad (3.4)$$

with $x_{ij}$ as the earlier reading measure for pupil $i$ in school $j$. The model differs from those used in Chapter 2 because it allows for variation in progress

## 3.4 Multilevel models for school and teacher effectiveness

between schools, via $u_j$ (or $b_{0j}$). It is this model which is the basic model used in studies of school and teacher effectiveness, although it can, of course, be used whenever there is a hierarchically structured two-level dataset, with a continuous response and a continuous explanatory variable. We represent it diagrammatically by Figure 3.3, which shows that there is a different intercept for each school (although, for now, we do not allow the slopes to vary from school to school).

We use the square root transformation of reading as $y$, as before, and we also 'centre' $x$ about its overall mean. In other words, we use:

$$x_{ij} = \tilde{x}_{ij} - \tilde{x}_{..} \qquad (3.5)$$

where $\tilde{x}_{ij}$ is the raw score and $\tilde{x}_{..}$ is the overall mean. This implies that the overall mean of $x_{ij}$ is zero.

The scatterplot of $y$ against $x$ (Figure 3.4) suggests a degree of non-linearity which we might want to represent by a quadratic term in $x$. Indeed, the coefficient for $x^2$, labelled $b_2$, does turn out to be important. (Note that coefficients for quadratic, cubic etc terms in $x$ are estimated more stably if $x$ is centered about its overall mean.) The results are:

$\hat{b}_0 = 3.5\ (0.066)$ – the intercept
$\hat{b}_1 = 0.96\ (0.036)$ – the coefficient for $x$
$\hat{b}_2 = -0.07\ (0.031)$ – the coefficient for $x^2$
$\hat{\sigma}_u^2 = 0.05\ (0.021)$ – between school variance in reading progress
$\hat{\sigma}_e^2 = 0.65\ (0.033)$ – between pupil within school variance

It is the random effects which particularly interest us here. If we compare the variance terms with those obtained for the model for attainment in Section 3.3, we see that the between school component has fallen by 69% and the between

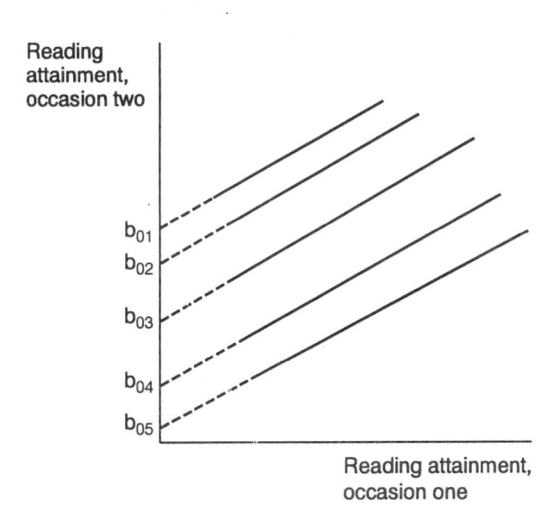

**Figure 3.3** Two-level model with random intercepts

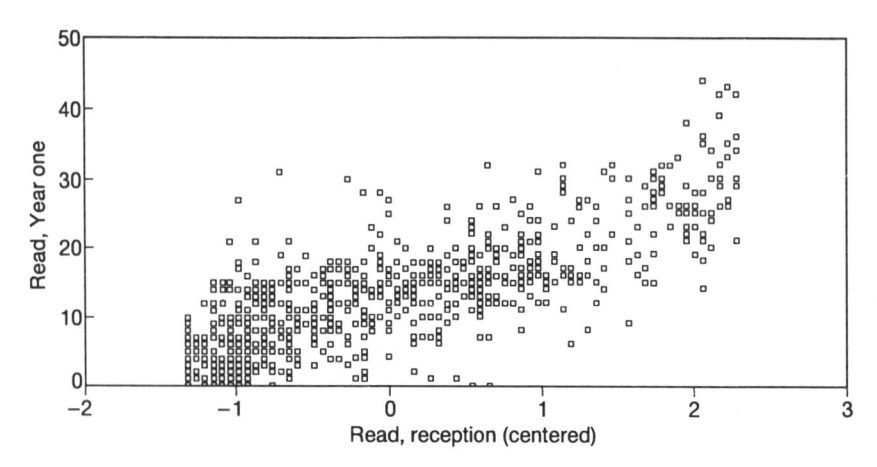

**Figure 3.4** Scatterplot of reading, Year one by reading, reception

pupil component has fallen by 54%. This is to be expected; for pupils, the best predictor of present attainment is past attainment, and for schools, some of the differences in attainment at one point in time are likely to be accounted for by earlier attainment differences. It is the existence of these earlier differences, often called intake differences, which make judgements about school and teacher differences based just on attainment so unsatisfactory (see Exercise 1). (Only if pupils were randomly assigned to schools might it be reasonable to judge their relative effectiveness on the basis of attainment.) Unfortunately, the so-called league tables of schools' performance published by the British government are, at present, based on attainment and not on progress. With these data, and this model, schools explain about 7% of variation in progress (compared with 10% of variation in attainment).

If we use classroom as our level two unit, we find $\hat{\sigma}_u^2 = 0.064$ (0.022) and $\hat{\sigma}_e^2 = 0.64$ (0.033) and, as for attainment, we find a little more variation in progress between classrooms than we do between schools. The estimates of the fixed parameters – $b_0$, $b_1$ and $b_2$ – are essentially the same in the two models.

Turning to the three-level model:

$$y_{ijk} = b_0 + b_1 x_{1ijk} + b_2 x_{1ijk}^2 + v_k + u_{jk} + e_{ijk} \qquad (3.6)$$

we get multilevel estimates, which are compared in Table 3.1 with the ordinary least squares (OLS) estimates for the fixed effects. The OLS estimates assume simple random sampling.

We see from Table 3.1 that, for progress, between school variance is a little higher than between classroom within school variance, but both exhibit considerable sampling variability. We also see that the OLS estimates are very similar to the multilevel estimates of the fixed effects, but the OLS standard errors are a little lower. However, the differences between the two sets of standard errors are not marked, mainly because the differences between schools and classrooms are small compared with the differences between pupils. If there had been more clustering, or similarity, of pupils within classrooms and schools, then there would have been more of a gap between the two sets of standard errors. In addition, we

## 3.4 Multilevel models for school and teacher effectiveness

**Table 3.1** Comparing multilevel and OLS estimates for reading progress

| Parameter | Multilevel Estimate | s.e. | OLS Estimate | s.e. |
|---|---|---|---|---|
| $b_0$ | 3.5 | 0.066 | 3.5 | 0.043 |
| $b_1$ | 0.96 | 0.036 | 0.96 | 0.035 |
| $b_2$ | −0.067 | 0.031 | −0.073 | 0.031 |
| $\sigma_u^2$ (School) | 0.04 | 0.023 | n.a. | |
| $\sigma_v^2$ (Classroom) | 0.023 | 0.018 | | |
| $\sigma_e^2$ (Pupil) | 0.64 | 0.033 | | |

must never forget that multiple regression estimated using OLS cannot tell us about the division of the residual variance into the three components of school, classroom and pupil. Instead, we just estimate a single residual variance, which is 0.70 here.

Again, we should check this model. Figure 3.5 shows the distribution of the estimated level three, or school, residuals. There is a suggestion of an outlier here, a school with a mean reading attainment much lower than expected on the basis of its pupils' initial reading level. Figure 3.6 gives the corresponding distribution for the estimated level two, or classroom, residuals, which appears well behaved, in the sense of being like a Normal distribution. When we look in a little more detail at the outlying school, we find that only one of the two classrooms in that school is at all far away from the mean. Moreover, this classroom does not appear to be an outlier. It is not always a straightforward task to assess the importance of outliers in a multilevel model, and the issue is discussed in some detail by Langford and Lewis (1998).

We must be extremely careful if we intend to use the rank order of the higher level residuals to provide a rank order of schools or teachers in terms of their effectiveness. These residuals are estimated from sample data and are therefore subject to sampling error. We can construct confidence intervals for each residual, and often these confidence intervals reveal a great deal of overlap between the individual schools. In other words, it is often impossible reliably to

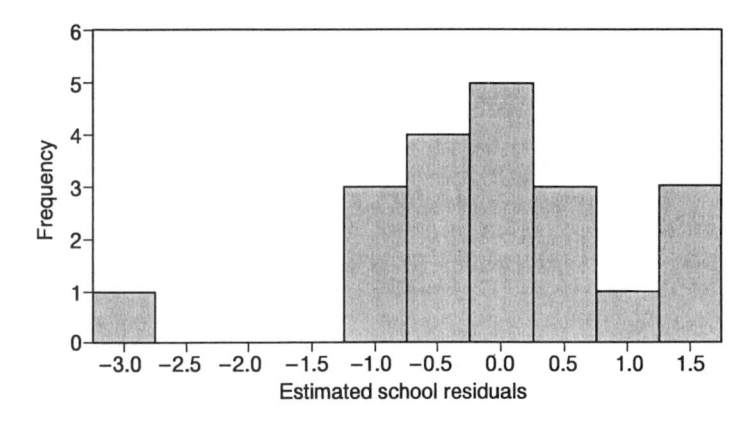

**Figure 3.5** Histogram of standardized school residuals

## 42  Populations with Structure

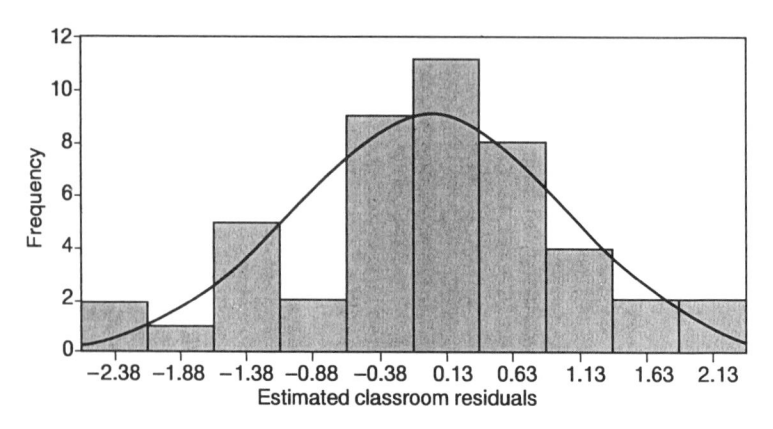

**Figure 3.6**  Histogram of standardized classroom residuals

distinguish the great majority of schools in terms of their effectiveness. This issue is examined in the broader context of ranking all kinds of institutions in a paper by Goldstein and Spiegelhalter (1996) and in the associated discussion.

### 3.5  Random intercepts and random slopes

Suppose that, rather than the picture in Figure 3.3, we have a situation like Figure 3.7. In other words, as well as the intercepts varying from school to school, the slopes linking $y$ to $x$ do too. Here, we are extending the interaction model of Section 2.7 in the sense that the relation between the response and the explanatory variable can vary by school. One of the strengths of the multilevel approach is its ability to deal with random slopes, and thus to be able to ask questions of data which could not be asked in the past. Note that we do *not* estimate a separate regression for each school. Instead, we write a two-level model with one explanatory variable, and random intercept and random slope, as follows:

$$y_{ij} = b_{0j} + b_{1j} x_{ij} + e_{ij} \tag{3.7}$$

We see that both the intercept and the slope now have a subscript $j$ to show that they can vary from school to school.

We can write:

$$b_{0j} = b_{00} + u_{0j} \tag{3.8}$$

and

$$b_{1j} = b_{10} + u_{1j} \tag{3.9}$$

There are now two level two variances to estimate – $\sigma_{u(0)}^2$ for the intercept and $\sigma_{u(1)}^2$ for the slope – and a level one variance as before. In addition, we must also include a covariance term in our model – $\sigma_{u(0)u(1)}$ – because, in general, intercepts and slopes will be correlated. In Figure 3.7, for example, we can see that the larger the intercept, the lower the slope, implying a negative correlation between

## 3.5 Random intercepts and random slopes

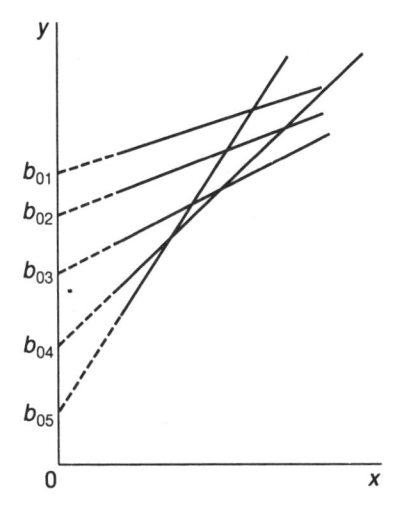

**Figure 3.7** Two-level model with random intercepts and random slopes

intercept and slope. We assume that the level two residuals, $u_{0j}$ and $u_{1j}$, are Normally distributed.

As soon as we allow for the possibility that slopes vary randomly, we are also introducing the possibility that there is no longer a single rank order of schools in terms of their effectiveness. Instead, some schools could be more effective for low attaining pupils whereas others could be more effective for high attainers. We can see this in Figure 3.7. If we compare school one with intercept $b_{01}$, and a relatively flat slope, with school five with intercept $b_{05}$ and a steep slope, we see that school one is more effective for low attainers (small x), whereas school five is more effective for high attainers (large x).

Let us illustrate these rather more complicated ideas with an example. We return to the data used in Chapter 2 on mathematics attainment and curriculum coverage, this time using the full dataset of 407 pupils in 24 classrooms (dataset 3.2 on the disk).

We start with a simple model for progress – mathematics attainment at the end of Year two as the response and mathematics attainment at the beginning of the year as the explanatory variable. Thus:

$$\text{MATH2}_{ij} = b_{0j} + b_1 \text{ZMATH1}_{ij} + e_{ij} \qquad (3.10)$$

We find the estimated between classroom variance is 4.1 (s.e. = 2.0) and the estimate of $\rho$, the intra unit correlation, is 0.09. We also find that the *deviance*, which is a measure of goodness of fit, is 2717.5 for this model. Now we allow the slope to be random so that:

$$\text{MATH2}_{ij} = b_{0j} + b_{1j} \text{ZMATH1}_{ij} + e_{ij} \qquad (3.11)$$

We have added a term for the random slope, and the covariance between slope and intercept, and the deviance falls to 2716.2. Hence, the difference between the two deviances is 1.3. The difference between the deviances for two 'nested' models, when one model includes the other model as a simpler case, has a $\chi^2$

distribution with degrees of freedom equal to the number of extra parameters estimated in the more complicated model. (Deviance is explained in more detail in Aitkin *et al.*, 1989.) Here, there are two degrees of freedom – corresponding to the variance and covariance – and so the difference in the deviances (1.3) does not approach statistical significance. This, in turn, means that the fit of the model is not improved by the addition of the two extra parameters. In other words, there is no evidence for random slopes for these data; differences in progress between classrooms are constant for all values of initial mathematics attainment, as illustrated in Figure 3.3.

Let us now extend the model we used in Section 2.2: mathematics attainment (MATH; $y_{ij}$) related to curriculum coverage (here, ZCURRIC; $x_{ij}$). The random intercepts model has a deviance of 2846.3. This deviance falls to 2837.1 when we move to a random slopes model – a reduction of 9.2 with two degrees of freedom ($p<0.01$). This means that, this time, a random slopes model does give a better fit to the data. The estimates of the random parameters are as follows:

$\hat{\sigma}^2_{u(0)}$ = 8.8 (3.8) – this is the random intercepts variance between classrooms

$\hat{\sigma}^2_{u(1)}$ = 4.0 (2.7) – this is the random slopes variance

$\hat{\sigma}_{u(0)u(1)}$ = −1.1 (2.3) – this is the covariance between intercept and slope

$\hat{\sigma}^2_e$ = 56 (4.1) – this is the between pupil within classroom variance.

We can use these estimates in two ways. First, we can calculate the correlation between intercept and slope which is just $-1.1/\sqrt{(8.8 \times 4)} = -0.19$. This tells us that there is a small negative correlation between intercept and slope for this model. (We should be a little wary of this statistic, for reasons which are explained in more detail in Chapter 4.) Second, we can see how the overall level two variance – in other words, the overall variance in mathematics attainment between classrooms – varies according to the value of ZCURRIC. It is equal to:

$$\hat{\sigma}^2_{u(0)} + 2\hat{\sigma}_{u(0)u(1)} \text{ZCURRIC} + \hat{\sigma}^2_{u(1)} \text{ZCURRIC}^2 \qquad (3.12)$$

so for ZCURRIC one standard deviation below its mean of zero, it is 8.8+2.2+4=15.0 and for ZCURRIC one standard deviation above the mean, it is 8.8−2.2+4=10.6. At the mean of ZCURRIC, it is just 8.8. We can see this graphically in Figure 3.8. The between classroom variance in attainment is smallest close to the mean of ZCURRIC and rises as we move away from the mean in both directions.

The histograms of the two sets of random effects, or estimated level two residuals, are shown in Figures 3.9 and 3.10. There is a suggestion in Figure 3.9 that a couple of classrooms have rather high intercepts, and similarly in Figure 3.10, one of the classrooms has a steeper slope than the others. However, by plotting intercepts against slopes as in Figure 3.11, we see that these high intercepts and steep slopes do not go together. These diagnostic plots do not reveal anything that would lead us to view this fitted model with any alarm.

So far in this chapter, all our model checking has focused on the higher level residuals. However, we should not forget the level one residuals, because we are making the same assumptions about them that we made in Chapter 2. Figure 3.12 shows the plot of the estimated standardized level one residuals against ZCURRIC. It appears that these residuals are heteroscedastic, just as they were for the sub-sample used in the example in Section 2.2. Hence, we might again

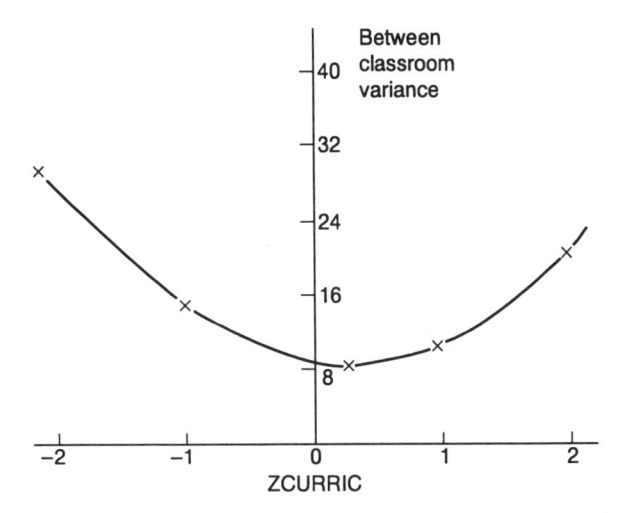

**Figure 3.8** Between classroom variance as a quadratic function of ZCURRIC

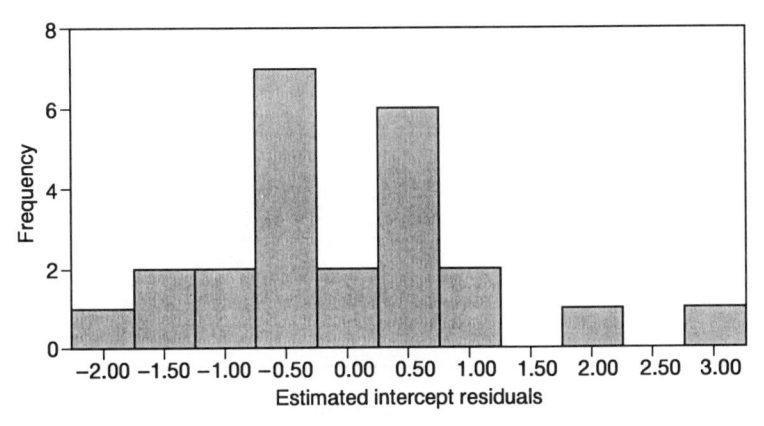

**Figure 3.9** Histogram of standardized intercept residuals

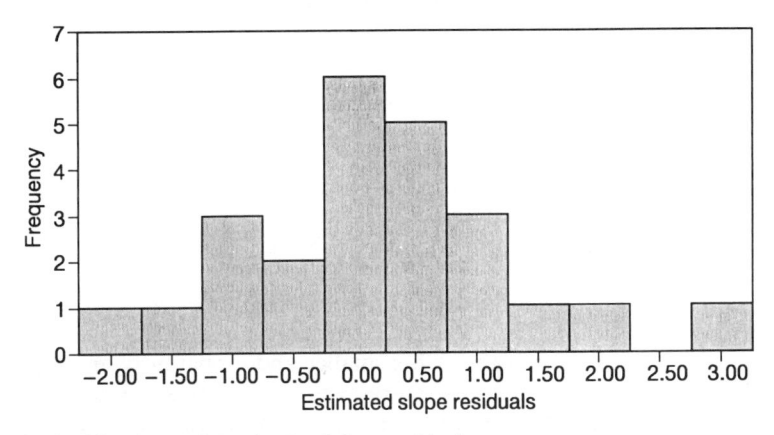

**Figure 3.10** Histogram of standardized slope residuals

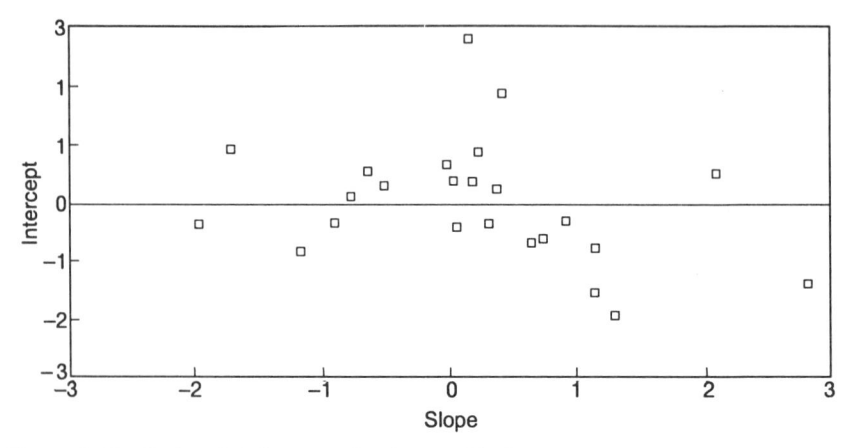

**Figure 3.11** Scatterplot of intercept by slope residuals

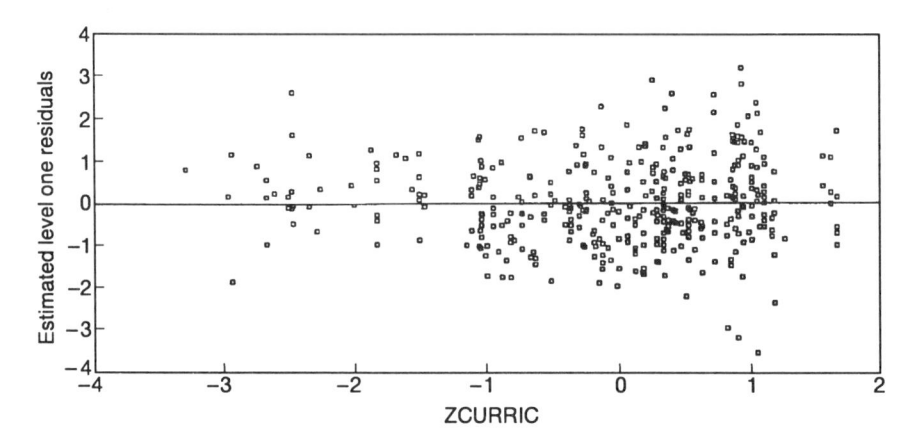

**Figure 3.12** Scatterplot of standardized level one residuals by ZCURRIC

improve our model by taking a log transformation of MATH2. In the next chapter, we will find another way of dealing with heteroscedasticity at level one, within the multilevel framework.

We should remember that both sets of estimated residuals, $\hat{u}_{0j}$ and $\hat{u}_{1j}$, are shrunken in the way described on p. 37. Hence, the $\hat{u}_{1j}$ show less variation than would the estimated slopes from separate OLS regressions fitted to the data for each school.

Let us now compare the fixed effects from this model with those obtained from a simple regression. Now we find rather substantial differences. The OLS estimate for ZCURRIC in Table 3.2 is smaller and, apparently, more precisely estimated. Not only is the simple regression less informative here, it is also misleading because it overestimates the precision of the estimate of ZCURRIC.

We could, in principle, go on to ask whether the introduction of the earlier

**Table 3.2** Comparing multilevel and OLS estimates of fixed effects

| Parameter | Multilevel | | OLS | |
|---|---|---|---|---|
| | Estimate | s.e. | Estimate | s.e. |
| CONSTANT | 29 | 0.74 | 29 | 0.40 |
| ZCURRIC | 8.2 | 0.65 | 6.9 | 0.40 |

measure of mathematics attainment changes the picture. The single level version of this model was fitted in Section 2.5. However, it is not now possible to estimate a model where the coefficient for ZCURRIC is random at level two because the computer algorithm does not converge. This is because the number of level two units is rather small, and perhaps also because, once ZMATH1 is introduced into the model, the variability between the ZCURRIC slopes is small. This illustrates one of the problems which users of multilevel techniques have to face. The method works best with large samples, especially at the highest level of the observed hierarchy. We need to construct large datasets in order fully to exploit the potential of multilevel modelling. Here we have just 24 classrooms, which is on the low side. We can satisfactorily estimate random intercepts models with this number of level two units, but, as we have just seen, it is sometimes more difficult to estimate random slopes models. The problem is not so different from those faced by users of multiple regression with small samples and, potentially, a large number of explanatory variables. We return to this issue, in the guise of statistical power, in Section 3.7.

The analyses in this section have given some indication about how we should specify and estimate multilevel models in an incremental way. In other words, we start with a simple model with just random intercepts, and then build on this by allowing slopes to be random, and including these additional parameters if the goodness of fit of the model to the data is significantly improved. We may well try a number of specifications before settling on what we believe is the most acceptable model for the data available to us, not forgetting the importance of model checking. This shows the creative side of model building, as we try to balance theoretical plausibility, explanatory power of the data available, and parsimony in statistical modelling. The issue of model selection is one we return to in the next chapter.

## 3.6 Explaining variability in intercepts and slopes

So far, we have only told half the story about multilevel modelling. We have not considered why intercepts and slopes vary across schools and classrooms. In other words, we do not yet have any tools to help us find out about the characteristics of more and less effective schools. In Section 3.4, we found out that pupils make more reading progress in some schools, and in some classrooms, than in others. In Section 3.5, we found that the relation between mathematics attainment and curriculum coverage varied across classrooms. These are interesting descriptions, but the models we used there do not provide us with any explanations. Consequently, at this point, we need to introduce the idea of level two, or higher level, explanatory variables. These are variables which vary from school to school, but not from pupil to pupil within schools. To give just two examples, the amount of funding received by a school, and the proportion of pupils receiving free school meals in a school, are both level two variables. In general, higher level variables can vary from unit to unit at that level, but are constant within each lower level.

Returning to the model in the previous section, we had:

$$y_{ij} = b_{0j} + b_{1j} x_{ij} + e_{ij} \tag{3.13}$$

and, because we did not have any level two variables, we also had:

$$b_{0j} = b_{00} + u_{0j} \tag{3.14}$$

and

$$b_{1j} = b_{10} + u_{1j} \tag{3.15}$$

where $b_{00}$ and $b_{10}$ are the fixed effects and $u_{0j}$ and $u_{1j}$ are the Normally distributed random effects. Let us now suppose we have two level two explanatory variables, $z_1$ which may account for variation in the intercepts, and $z_2$ which may account for variation in the slopes ($z_1$ and $z_2$ could, of course, be the same variable). So we now write:

$$b_{0j} = b_{00} + b_{01} z_{1j} + u_{0j} \quad \text{for the random intercepts} \tag{3.16}$$

and

$$b_{1j} = b_{10} + b_{11} z_{2j} + u_{1j} \quad \text{for the random slopes} \tag{3.17}$$

where $z_1$ and $z_2$ carry the subscript $j$ to show that they vary from school to school, but not $i$ indicating that they are constant from pupil to pupil within a school. There are two points to note about these extensions to the model. The first is that there is little point to them if initial analyses suggest that there is no more variation in intercepts or slopes than would be expected by chance. The second is that, if $z_1$ or $z_2$ or both are important explanatory variables at level two, then their inclusion will have the effect of reducing the level two variances, $\sigma^2_{u(0)}$ and $\sigma^2_{u(1)}$, possibly close to zero.

If we substitute the equations for $b_{0j}$ and $b_{1j}$ into the original model, we have:

$$y_{ij} = b_{00} + b_{01} z_{1j} + u_{0j} + (b_{10} + b_{11} z_{2j} + u_{1j}) x_{ij} + e_{ij} \tag{3.18}$$

or

$$y_{ij} = b_{00} + b_{01} z_{1j} + b_{10} x_{ij} + b_{11} x_{ij} z_{2j} + u_{0j} + u_{1j} x_{ij} + e_{ij} \tag{3.19}$$

This looks rather a daunting model but, in fact, the ideas it contains are no more complicated than those presented in Chapter 2. We see that the level two variable, which accounts for some of the variability between schools in the intercept ($z_{1j}$), is just one of the terms of the expanded model. However, $z_{2j}$ – the variable accounting for some of the variation in the slopes – appears as an interaction term ($x_{ij} z_{2j}$) and this interaction is sometimes called a *cross-level interaction*. We also see that the random slope term, $u_{1j}$, appears as an interaction with $x_{ij}$, which is why the level two variance is not constant, but varies with $x$.

Many school effectiveness studies will collect extensive data at level two, and will therefore have available many potential $z$-type variables such as the experience and training of the headteachers, and the condition of the school buildings.

## 3.6 Explaining variability in intercepts and slopes

However, some studies of pupils' progress do not collect a lot of data at higher levels. Nevertheless, this absence of data at higher levels does not mean that there is no scope for explaining level two variability. We can generate level two explanatory variables from the level one data. In particular, we have available the number of level one units for each level two unit, and this variable can sometimes be used as a proxy for class or school size. Also, we can calculate aggregate statistics for each level two unit such as the mean and standard deviation of initial attainment. We can also calculate proportions – the proportion of boys in a school, for example. These aggregate statistics are often called *contextual* variables.

Let us now return to the example used in Section 3.5, where mathematics attainment was related to curriculum coverage (dataset 3.2). We will use the number of pupils in each classroom, and the mean curriculum coverage for each classroom, as our level two, or $z$, variables. Let us start with the number of pupils in the classroom; this is a proxy for class size but is measured with error because of missing data. We have:

$$\text{MATH2}_{ij} = b_{0j} + b_{1j} \text{ZCURRIC}_{ij} + e_{ij} \qquad (3.20)$$

$$b_{0j} = b_{00} + b_{01} \text{SIZE}_j + u_{0j} \qquad (3.21)$$

$$b_{1j} = b_{10} + b_{11} \text{SIZE}_j + u_{1j} \qquad (3.22)$$

When we just include class size on its own in the model (i.e. to explain $b_{0j}$ but not $b_{1j}$), we find essentially no change in the deviance. This suggests that class size cannot account for any of the variance between classrooms in mean mathematics attainment, after allowing for curriculum coverage. We then add the interaction between class size and curriculum coverage to see whether class size explains any of the variation in the relation between attainment and coverage across classrooms (i.e. to explain $b_{1j}$). There is a small improvement in the fit of the model; there is a fall of 2.25 in the deviance (see p. 43) which, for one degree of freedom, gives a *p*-value of 0.14. The variability in the slopes falls by about 12%. The fixed effects are:

| | |
|---|---|
| Constant, i.e. $\hat{b}_{00}$ | = 29 (2.9) |
| ZCURRIC, i.e $\hat{b}_{10}$ | =  4.1 (2.8) |
| SIZE, i.e. $\hat{b}_{01}$ | = –0.037 (0.16) |
| SIZE by ZCURRIC, i.e. $\hat{b}_{11}$ = | 0.24 (0.16) |

What we see is that the effect of curriculum coverage on attainment ($b_{10}$) appears, if anything, to become stronger as class size increases, because $\hat{b}_{11}$ is positive.

We now look at the effect of mean curriculum coverage: do pupils have higher mathematics attainment if they are taught in classrooms where more of the curriculum is covered, regardless of how much they cover themselves? In order to answer this question satisfactorily, we should make a slight adjustment to our model. Rather than use ZCURRIC, we 'centre' the curriculum coverage scores around each classroom's mean so that deviations from the mean *within* each classroom sum to zero. We can then more easily separate the within classroom effect from the between classroom effect. The issue of 'centering' is a somewhat controversial one in multilevel modelling, but most researchers now agree that there is no one solution to the problem. Rather, decisions about whether, and how, to centre depend on the substantive question to be answered. For a full discussion of 'centering', refer to Kreft *et al.* (1995). To keep things simple, we confine ourselves just to a random intercepts model, i.e.:

$$\text{MATH2}_{ij} = b_{0j} + b_1 (\tilde{x}_{ij} - \bar{x}_j) + e_{ij} \tag{3.23}$$

$$b_{0j} = b_{00} + b_{01} \bar{x}_j + u_{0j} \tag{3.24}$$

with $\tilde{x}_{ij}$ as the curriculum coverage score (in this example, the raw score) and $\bar{x}_j$ mean curriculum coverage in a classroom.

These two equations, when combined, become:

$$\text{MATH2}_{ij} = b_{00} + b_{01} \bar{x}_j + b_1 (\tilde{x}_{ij} - \bar{x}_j) + u_{0j} + e_{ij} \tag{3.25}$$

so mean curriculum coverage $\bar{x}_j$ is $z_{1j}$, and there is no interaction term because we are not allowing slopes to be random. Before including mean curriculum coverage, our model has a between classroom variance for the intercept of 19 (s.e. = 6.6). After including mean curriculum coverage, this falls by two thirds to 6.2 (s.e. = 2.9). The fall in deviance is 19.4 (1 df), which is highly statistically significant. The fixed effects are:

Constant, i.e. $\hat{b}_{00}$ = 9.7(3.6)
Centered curriculum coverage, i.e. $\hat{b}_1$ = 0.088 (0.0057)
Class mean coverage, i.e. $\hat{b}_{01}$ = 0.069 (0.012)

We see that the two coverage effects are similar in size. A unit change in curriculum coverage within a class, and a unit change in mean curriculum coverage, both have marked, and roughly equal, effects on attainment. The situation we are describing here is akin to the one represented diagrammatically in Figure 2.13, although there, the within and between group regressions had opposite signs. More details of analyses similar to this one can be found in Plewis (1996a).

Clearly, there is scope for extending the complexity of these models. Some of these extensions are presented in the next chapter, some are briefly covered in Chapter 8. But, for the final main section of this chapter, we turn to an important issue in educational research – the design and analysis of intervention studies.

## 3.7 Multilevel models for intervention studies

An important part of quantitative educational research is to evaluate the effectiveness of those educational interventions such as Reading Recovery (see Exercise 4) which aim to affect pupils' performance in some way. However, analyses of these sorts of studies often fail to reflect the multilevel nature of the data. Also, they ignore differences between those studies in which the intervention is allocated at the level of the classroom or school, and studies where the intervention is allocated at the level of the pupil. The level at which the intervention is assigned should influence the way in which data from the study are analysed.

## 3.7 Multilevel models for intervention studies

For simplicity, we assume a basic design in which there are two groups – intervention and control – and two levels – classroom (or teacher) and pupil. We will also assume that the intervention has been assigned randomly, and we have a pre-intervention measure of the outcome. We do not meet any new modelling ideas in this section, rather we apply the ideas of the previous four sections to this particular problem.

First, we consider designs in which the intervention is assigned at the level of the teacher (level two). This could be some form of in-service training, for example. Our basic model, for a response or post-test $y$, pupil attainment say, is then:

$$y_{ij} = b_{0j} + b_1 x_{ij} + e_{ij} \qquad (3.26)$$

where $x$ is a pre-test measure of attainment, which we will assume has a sample mean of zero. This is just the first model of Section 3.4.

Let us assume that there is some variation in the mean level of attainment across classrooms, after allowing for the effects of the pre-test. In other words, $b_{0j}$ does not equal $b_0$ for all $j$. We want to know if any of this level two variation can be explained by the binary variable defining the intervention, i.e.:

$$b_{0j} = b_{00} + b_{01} z_{1j} + u_{0j} \qquad (3.27)$$

where $z_{1j} = 1$ if intervention and 0 if control and $z_{1j}$ is a level two variable, as explained in Section 3.6. (It does not vary from pupil to pupil within a classroom.) Is $\hat{b}_{01}$, the estimated intervention effect, greater than we would expect by chance? Note that any analysis which ignored the clustering of pupils within classrooms, as any single level ordinary least squares analysis would do, will give an over-precise estimate of the intervention effect.

Up to now, we have assumed that the relation between the response and the pre-test is constant across classrooms. If we relax this assumption, as we did in Section 3.5, we replace $b_1$ by $b_{1j}$ and make $b_1$ random at level two. Can any of this level two variation in the slopes be explained by the intervention variable? In other words, is the estimate of $b_{11}$ in:

$$b_{1j} = b_{10} + b_{11} z_{1j} + u_{1j} \qquad (3.28)$$

greater than we would expect by chance? This is equivalent to asking if the intervention is differentially effective according to initial attainment (or any other level one explanatory variable), because if we substitute $b_{0j}$ and $b_{1j}$ into the basic model, we end up, essentially as we did in Section 3.5, with:

$$y_{ij} = b_{00} + b_{01} z_{1j} + b_{10} x_{ij} + b_{11} x_{ij} z_{1j} + u_{0j} + u_{1j} x_{ij} + e_{ij} \qquad (3.29)$$

and so $b_{11}$ is the coefficient of the cross-level interaction term between the pre-test and group, $x_{ij} z_{1j}$.

If there is a positive intervention effect ($\hat{b}_{01} > 0$), and if $\hat{b}_{11}$ is also positive, then the intervention effect becomes stronger as the value of the pre-test increases. If $\hat{b}_{11}$ is negative then the intervention effect becomes smaller than $\hat{b}_{01}$ as the pre-test increases from its mean of zero.

Assigning interventions at the classroom level has some advantages but it has one serious drawback. Only with a substantial number of classrooms in the study is it likely that we will be able to observe statistically significant intervention effects, both overall and, especially, those which vary with an explanatory

## 52  Populations with Structure

variable. Because there is no variation in $z_1$ within classrooms, the standard errors of $\hat{b}_{01}$ and $\hat{b}_{11}$ will be inversely proportional to $J$, the number of classrooms, and often $J$ is rather small. In other words, these designs tend to lack statistical power to pick up intervention effects, particularly if these effects are not very large.

Suppose assignment is at the level of the pupil (level one), as, for example, it would be if the intervention was extra tutoring of some kind, given to some pupils but not others within a classroom. Our model now is:

$$y_{ij} = b_{0j} + b_1 x_{ij} + b_2 z_{1ij} + e_{ij} \qquad (3.30)$$

Note that our intervention variable, $z_1$, now has a subscript $i$ as well as $j$, to show that it varies from pupil to pupil within a classroom. We want to know whether $b_2$ is greater than zero, but we must continue to take account of the structure of our data. Hence, we need to include $b_{0j}$ rather than just $b_0$ which is what we had with just the single level model of Section 2.7. The standard error for $\hat{b}_2$ will be inversely proportional to a number greater than $J$, the number of classrooms, but less than the number of pupils. Clearly, the intervention effect will be estimated more precisely with this design, the increase in precision over a design where assignment is at level two depending on the intra unit correlations for the pre- and post-tests. The smaller the intra unit correlations, the greater the increase in precision (Goldstein, 1995, p. 25).

We can extend the model by allowing not only $b_1$ but also $b_2$ to vary across classrooms, thus replacing $b_1$ and $b_2$ by $b_{1j}$ and $b_{2j}$. By allowing $b_2$ to be random at level two and so making the effect of the intervention random at level two, we introduce the possibility that the intervention is more effective in some classrooms than in others. In turn, we can try to explain this variation by variables defined at the classroom level, such as a measure of the teacher's enthusiasm for the content of the intervention. Thus, we can write:

$$b_{2j} = b_{20} + b_{21} z_{2j} + u_{2j} \qquad (3.31)$$

where $z_{2j}$ is, for example, an 'enthusiasm' measure.

Analyses of this kind are much more easily done with assignment at the pupil level. With assignment at the classroom level, the inclusion of an 'enthusiasm' measure together with the intervention variable in the equation for $b_{0j}$ is likely to introduce problems of collinearity (see p. 21), because the enthusiasm measure is bound to be zero for all the teachers in the control group.

The issue of whether the intervention is differentially effective according to the value of the pre-test (or any other explanatory variable) is easily answered by including an interaction term, $z_{1ij}x_{ij}$, in the model. These are both pupil level variables, and therefore the effect will be estimated more precisely than it would in a level two design. We can go further and allow the coefficient for the interaction to vary across classrooms:

$$y_{ij} = b_{0j} + b_{1j} x_{ij} + b_{2j} z_{1ij} + b_{3j} z_{1ij} x_{ij} + e_{ij} \qquad (3.32)$$

There is now the potential for explaining the variability across teachers in both $b_{2j}$ and $b_{3j}$ by teacher variables.

Whatever design is adopted to test the effectiveness of an educational intervention, this section has shown that the analysis of the data so generated needs to be located within a multilevel framework. Not only is this important in order to get an adequate representation of the precision of any intervention effect, it also offers various possibilities for investigating how the intervention effect operates, which go beyond a simple comparison of means across the two groups. As we have seen, we can examine whether the intervention effect is constant, or, importantly, whether it varies according to the value of background variables, measured both at the pupil level and at the classroom (or school) level. Moreover, we can also test whether the intervention affects variability as well as overall level. A more detailed discussion of these methodological issues, linked to practical issues from actual intervention studies, can be found in Plewis and Hurry (1997).

## 3.8 Software for multilevel modelling

There are two main specialist packages for multilevel modelling on a personal computer. These are HLM, Version 4, which operates in a Windows environment, and MLn, which will do so soon. HLM, produced by Bryk *et al.* (1996), is reviewed by Jones (1996). MLn, produced by the Multilevel Models Project at the Institute of Education, University of London (Rasbash and Woodhouse, 1995), is reviewed by Wright (1996), and was used for the examples in this chapter and the next. In addition, some multilevel modelling is possible within the major statistical packages such as SAS (PROC MIXED). Rapid developments in statistical computing, and methodological advances in modelling make further developments in software likely.

We have now reached the end of our introduction to multilevel models. The next chapter continues the multilevel theme, applying it to repeated measures data.

## Exercises

Answers to those exercises marked '#' can be found at the end of the book. Exercises 1 and 3 ideally require access to a specialist multilevel modelling package or procedure.

1  Using dataset 3.1 on the disk, compare the rank orders of the residuals from three-level models for reading attainment and progress, for both schools and classrooms. What implications do these findings have for 'league tables' of institutions' performance.
(#)

2  Write down a two-level model for academic progress, with pupil as level one and school as level two, with random intercepts and random slopes. Draw a sketch to illustrate a positive correlation between intercepts and slopes and consider the implications of this positive correlation for school effectiveness research.

3  Using dataset 3.2 on the disk, obtain between and within classroom variances for curriculum coverage, both before and after allowing for initial mathematics attainment. What are the implications of these findings?
(#)

**4** An evaluation of the Reading Recovery programme (Sylva *et al.*, 1995) had the following design. Pupils receiving the Reading Recovery programme were compared both with control pupils within the same classrooms, and with pupils in schools not receiving an intervention. Set out a plan for the analysis of the data generated by this design, which includes the relevant statistical models.

# 4
# Growing and Changing: Repeated Measures of Continuous Responses

## 4.1 Introduction

In Chapter 3, we saw how the hierarchical structure of educational data, with pupils nested in classrooms, which in turn are nested in schools, should influence statistical analysis. In this chapter, we consider another kind of structure that is often found in educational research. Here, we are concerned with data that are generated by repeated measurement of the same units, and are therefore structured by time (or age). Longitudinal datasets of this type have a lot in common with those discussed in the previous chapter, but there are also, as we shall see, some subtle differences.

This chapter starts by illustrating the hierarchical nature of repeated measures data, and showing why a multilevel approach to their analysis has several advantages over other methods. The bulk of the chapter is devoted to an analysis of a set of repeated measures of reading attainment, which is used to show how to specify, estimate and interpret some basic two- and three-level growth curve models, and how to deal with missing data. These ideas are then extended to take account of important issues about the scale of the response. The chapter ends with a comparison of two contrasting approaches to the analysis of longitudinal data. As before, the emphasis of the chapter is on model specification and the interpretation of results, and not on technical issues of computation and estimation.

## 4.2 The hierarchical nature of repeated measures data

Suppose we have a sample of pupils whose reading attainment is measured on a number of occasions. At their simplest, we can think of these data as having a hierarchical or nested structure with two levels, as shown in Figure 4.1. The pupils define level two, the repeated measures or occasions define level one. The important difference between Figure 4.1 and Figure 3.1 is the position of pupils within the hierarchy; in Figure 3.1 they were at level one but in Figure 4.1 they

56   *Growing and Changing*

are at level two. Another practical difference is that, with longitudinal designs, we usually have a large number of level two units and rather few level one units, the reverse of the designs we considered in Chapter 3. We can, of course, extend Figure 4.1 to include higher levels representing the fact that the pupils will be in classrooms in schools. It is also worth bearing in mind that our repeated measures could be of schools or teachers rather than (or even as well as) of pupils. So we might have a four-level structure like the one illustrated in Figure 4.2, with a sample of schools, studied over time by measuring repeated cohorts of pupils, these pupils themselves being repeatedly measured as they pass through the school. This is clearly likely to be a large study, but one with great potential for looking at the stability of school effects as well as for studying pupils' educational growth (see Exercise 4).

In the previous chapter, we did not require balanced data. In other words, it was not necessary to have the same number of lower level units in each higher level unit. In the same way here, we do not need each pupil to have the same number of measurements. This is one of the great advantages of a multilevel approach to repeated measures analysis. Often, in longitudinal studies, pupils are lost from the study entirely, perhaps because they move school. It is also possible for a pupil to miss one measurement, perhaps because they were absent from school, but to be measured on subsequent occasions. All the data which are available should be used in any statistical analysis and, within a multilevel model, they can be, even from pupils with just one measurement. This is in marked contrast to analyses based on repeated measures analysis of variance

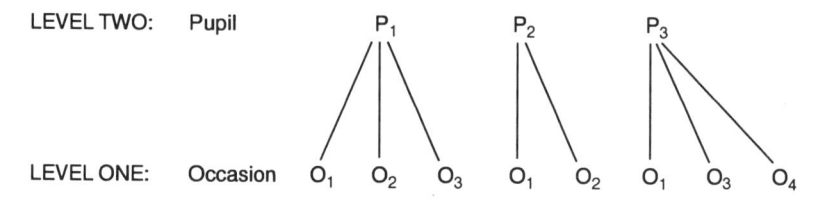

**Figure 4.1**   The hierarchical nature of repeated measures data

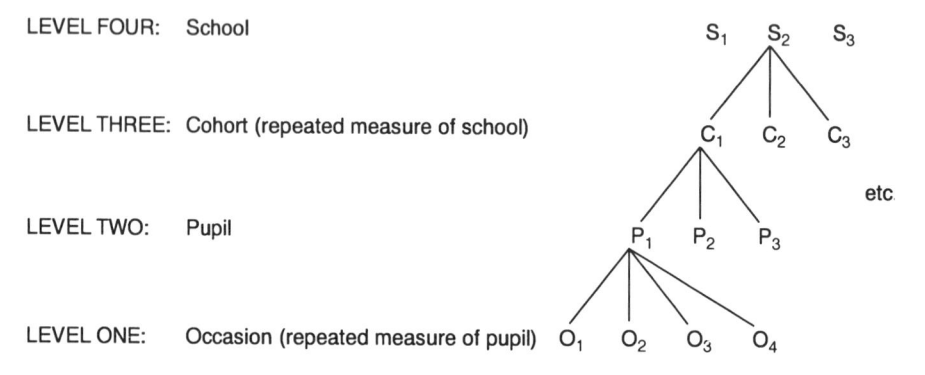

**Figure 4.2**   A structure with four levels and two repeated measures

and multivariate analysis of variance (MANOVA), the traditional approaches to the analysis of this sort of data (see, for example, Bock, 1979). Not only can these traditional methods be rather difficult for beginners to understand, they can usually only be applied to pupils with complete data. For example, the MANOVA procedure in SPSS can only be used with complete data. As we shall see in Section 4.6, this can result in analyses based on a rather small proportion of the available data, which is both inefficient and which can also introduce biases.

At this point, we meet the data to be used to illustrate the methods (dataset 4.1 on the disk). They come from a longitudinal study of a cohort of pupils who entered 33 multi-ethnic inner London infant schools in 1982, and who were followed until the end of their junior schooling in 1989. More details about the study can be found in Tizard *et al.* (1988) and Plewis (1991b). Pupils' reading abilities were tested on up to six occasions; annually from 1982 to 1986 and in 1989. Reading attainment is our response, and we have three levels of data – school (level 3), pupil (level 2) and measurement occasion (level 1). In addition, we have three explanatory variables. The first is the pupil's age, which varies from occasion to occasion and is therefore a level one variable. The other two are sex and ethnic group, which vary from pupil to pupil and are thus level two variables. Our basic questions are essentially those raised in Chapter 1 (p. 7) and illustrated by Figure 1.2. They are as follows:

(i) How does reading ability develop as pupils get older?
(ii) Does this development vary from pupil to pupil and from school to school?
(iii) And, if so, does it vary systematically from one type of pupil (for example, girls) to another (boys), and according to characteristics of the school?

The initial sample at school entry was 277 – 171 white indigenous pupils and 106 black British pupils of African Caribbean origin. This rose to 371 one year later and fell to 198 by the end of junior school. Table 4.1 gives the number of reading tests per pupil and shows that only a minority of pupils were measured at every occasion (Plewis, 1996b). Altogether, 1758 observations were obtained on 407 pupils, of whom 259 were white and 148 were black. It is important to note that pupils with, say, a total of three tests did not necessarily all have tests at the same three occasions. Table 4.2 illustrates how the pattern of response might vary across pupils; pupil one had all six tests, pupil two had three tests – at occasions one, two and three, pupil three also had three tests but at occasions one, two and

**Table 4.1** Number of tests per pupil

| Number of tests | Number of pupils | % of total pupils | Inverse cumulative % |
| --- | --- | --- | --- |
| 1 | 37 | 9 | 100 |
| 2 | 41 | 10 | 91 |
| 3 | 42 | 10 | 81 |
| 4 | 48 | 12 | 71 |
| 5 | 113 | 28 | 59 |
| 6 | 126 | 31 | 31 |
| Total | 407 | 100 | n.a. |

**Table 4.2**  Different patterns of response and age at testing

| Pupil | Occasion |     |     |     |     |      |
|-------|----------|-----|-----|-----|-----|------|
|       | 1        | 2   | 3   | 4   | 5   | 6    |
| One   | 4.6      | 5.7 | 6.7 | 7.7 | 8.7 | 11.6 |
| Two   | 4.8      | 5.7 | 6.8 | –   | –   | –    |
| Three | 4.8      | 5.8 | –   | 7.8 | –   | –    |
| Four  | 4.9      | –   | –   | –   | –   | –    |

four, and pupil four had just one test, and so on. Table 4.2 also indicates that pupils' ages can vary at fixed measurement occasions – compare, for example, pupil one and pupil four at occasion one. This brings us to a second great advantage of multilevel modelling of repeated measures. This is the ability to distinguish age (or time) from occasion, and hence to be able to model them separately. Therefore, it is not necessary for all measurements to take place at exactly the same age or time at each occasion, nor for measurements to be evenly spaced.

One problem we have with data of this kind is just how we define and construct our response over age. Reading attainment cannot usually be measured with the same test at each age, and in our example four rather different, age-appropriate tests were used. The underlying, or latent, variable is clearly reading ability for all ages but the observed variables change with age. This means we do not have a fixed scale which covers the age range. Therefore, we have to construct a scale with sensible properties, at the same time recognizing that our chosen scale is bound to be arbitrary. Moreover, we may find that our results vary as we change the scale of our response. We return to this problem in Section 4.7, but, for now, we work with a scale for the response defined in the following 'age-equivalent' way. The mean reading score at each occasion is set equal to the mean pupil age for that occasion, and the variance is set to increase from one occasion to the next in such a way that the coefficient of variation (i.e the standard deviation divided by the mean) is constant and equal to 0.13. (This value comes from data presented by Bayley, 1949.) Allowing the variance, as well as the mean, to increase with age is consistent with what we know about growth of all kinds, and this measure has more face validity than an alternative approach of working with $z$ scores (zero mean and unit variance) at each occasion. Table 4.3 gives basic descriptive data for reading attainment and age across occasions. Note that age is measured as deviations from the grand mean and is therefore centered (see Section 3.4).

We now move on to see how we can set up some relatively simple models to answer our three questions about reading growth. All the models are estimated

**Table 4.3**  Means and SDs for reading attainment and age

| Occasion | Reading     | Age          |
|----------|-------------|--------------|
| 1        | 4.72 (0.61) | −2.41 (0.15) |
| 2        | 5.69 (0.74) | −1.44 (0.15) |
| 3        | 6.67 (0.87) | −0.46 (0.16) |
| 4        | 7.62 (0.99) | 0.49 (0.13)  |
| 5        | 8.61 (1.12) | 1.48 (0.13)  |
| 6        | 11.3 (1.47) | 4.21 (0.12)  |

using the package MLn, also used in Chapter 3. Readers interested in seeing how to construct the relevant MLn commands are referred to Plewis (1995), from which some of the results in this chapter are taken.

## 4.3 Basic multilevel growth curve models

We start with just a two-level model – pupils and occasions (repeated measures). Also, we restrict ourselves initially to linear, or straight line, growth which we assume to be the same for each member of the sample. Hence, our model, which is simplistic but serves as a baseline, is:

$$\text{READ}_{ij} = b_{0j} + b_1 \text{AGE}_{ij} + e_{ij} \qquad (4.1)$$

where $i$ ($i = 1$ ... maximum of 6) represents occasions and $j$ ($j = 1$ ... 407) represents pupils. Note that $j$ is the subscript for pupil because pupil now defines level two. In Chapter 3, subscript $i$ was used for pupil, where pupil defines level one. Our model allows mean reading ability to vary across pupils ($b_{0j}$ rather than $b_0$), but all pupils' reading grows at the same (constant) rate, $b_1$. This model is often referred to as a one way repeated measures analysis of variance with linear trend, and could be used in the same way with time rather than age defining the longitudinal structure.

The first point to make about this model is that, with these data, the estimate of $b_1$ is bound to be equal to one because of the way we have defined our scale for the response as an age-equivalent scale. (In Table 4.3, the differences in mean reading score are essentially one over the first five occasions, and close to three between occasions five and six, which were three years apart.) So all that is left is to see how the total variance divides into two components: between pupils (level two) and between occasions within pupils (level one). We find that two-thirds of the total is between pupils, one-third within pupils. So the level two variance is considerably greater than the level one variance, a common finding with repeated measures data but in sharp contrast to the findings from the models of school and teacher effectiveness, described in Chapter 3.

At the very least, we expect pupils to vary in their growth rates and so

$$\text{READ}_{ij} = b_{0j} + b_{1j} \text{AGE}_{ij} + e_{ij} \qquad (4.2)$$

with $b_{1j}$ replacing $b_1$, is more realistic. This gives us three level two random parameters (as in Section 3.5) – variance between pupils in reading at mean age ($\sigma^2_{u(0)}$), variance between pupils in growth rates ($\sigma^2_{u(1)}$), and a covariance term between reading level and reading growth rate, $\sigma_{u(0)u(1)}$. This model also implies that the variation between pupils is no longer constant but instead varies with age, which, of course, we know from the way we have defined our scale for the response (on p. 58). The estimates are:

$\hat{\sigma}^2_{u(0)}$ = 0.68 (s.e. = 0.053) – intercept variance across pupils
$\hat{\sigma}^2_{u(1)}$ = 0.037 (0.0039) – slope variance across pupils
$\hat{\sigma}_{u(0)u(1)}$ = 0.12 (0.012) – covariance between intercept and slope

This extended model fits much better because there is a reduction of 586 in the deviance for just two extra parameters (see Section 3.5 for a discussion of deviance). Pupils vary in their linear growth rates; the variance is 0.037 so the standard deviation is 0.19. Assuming a Normal distribution for these growth rates, then we would expect about 95% of them to lie within two SDs of the mean, which we know is equal to one here. In other words, the 95% coverage interval for the pupil growth rates is 0.62 to 1.38, which we can also interpret as saying that those pupils with the slowest growth rates gain no more than 0.6 years in reading in one calendar year, those with the fastest at least 1.4 years.

As always, we should check the residuals, and when we do, we find there is nothing untoward about them. We can also calculate the correlation between growth rate and reading level when (centered) age is zero – it turns out to be 0.77 showing the higher the reading level the faster the growth. This statistic has been given some importance in psychological theories about development. However, we should be wary of it. First, it depends on growth being appropriately represented by a straight line and, secondly, as we shall see in Section 4.8, it varies according to the scale adopted for the response.

It is somewhat unlikely that reading growth is linear over as wide an age range as the seven years we have here. Consequently, we should introduce some curvature by trying a quadratic term in age, which we allow to vary from pupil to pupil. (Note that the fixed parameter for the quadratic term in age – in other words, its mean value – must be zero, again because of the way the scale for the response is defined.) Our random part of the model at level two is now quite complex – there are three variances – for intercept, linear growth and the quadratic term – and three covariances. However, as these estimates are based on the number of pupils (i.e. 407), we estimate them with a reasonable degree of precision.

Our model is now:

$$\text{READ}_{ij} = b_{0j} + b_{1j}\,\text{AGE}_{ij} + b_{2j}\,\text{AGE}_{ij}^2 + e_{ij} \qquad (4.3)$$

One of the advantages of centering age is that we get more precise estimates when we introduce a quadratic age term. In principle, we could also introduce cubic and other terms in age, but this would be rather unwise here as we have at most six measurements per pupil.

Is this model an improvement over the model with just linear growth? The deviance falls by 61.8 for four extra parameters, much more than we would expect by chance. The random parameters are estimated to be:

$\hat{\sigma}^2_{u(0)} = 0.77$ (s.e. = 0.06) – this is the variance for reading level

$\hat{\sigma}^2_{u(1)} = 0.039$ (0.0042) – this is the variance for the linear term in age

$\hat{\sigma}_{u(0)u(1)} = 0.14$ (0.014) – this is the covariance between level and linear growth

$\hat{\sigma}^2_{u(2)} = 0.0013$ (0.00025) – this is the variance for the quadratic term in age

$\hat{\sigma}_{u(0)u(2)} = -0.014$ (0.0031) – this is the covariance between level and quadratic growth

$\hat{\sigma}_{u(1)u(2)} = -0.0020$ (0.00075) – this is the covariance between linear and quadratic growth.

$\hat{\sigma}^2_e = 0.13$ (0.0066) – this is the level one variance

The first three random parameters are not very different from those found for the linear growth model. The variance for the quadratic term in age is equivalent to a standard deviation of 0.036, so the 95% coverage interval for the coefficient of $AGE^2$ goes from $-0.072$ to $+0.072$. This implies that some pupils' reading ability grows more and more rapidly as they get older (those with a positive quadratic term) whereas, for others, it slows down (those with a negative quadratic term). We can also use the covariances to derive a set of correlations as follows:

correlation between level and linear growth = 0.80;
correlation between level and quadratic growth = $-0.45$;
correlation between linear and quadratic growth = $-0.28$.

Therefore, *for this scale*, on average, the higher the level the faster the linear growth but the lower the quadratic term, and the faster the linear growth the smaller the quadratic term.

We seem to have a reasonable model for growth, although we must not forget that it is a model for a particular, and arbitrary, scale. We have not, however, taken account of all the structure in our data. There are questions we should ask about differences between schools, both in reading level and in growth. On average, do some schools have higher reading levels than others and, more interestingly, do pupils get on faster in some schools than in others?

When we introduce school as a random effect at level three, the fit improves by 11.8 for this extra parameter ($p<0.001$) and the variance term is 0.037 (s.e. = 0.017). The random parameters at level two are little changed from the earlier model. If we allow the linear, but not the quadratic, growth rates to vary from school to school, the model is:

$$READ_{ijk} = b_{0jk} + b_{1jk} AGE_{ijk} + b_{2j} AGE^2_{ijk} + e_{ijk} \qquad (4.4)$$

where $k$ is the subscript for school, as in Chapter 3.

The addition of two more random parameters at level three to allow for the linear growth rates to vary across schools improves the fit by a further 5.7 ($p<0.06$). The estimates are:

$\hat{\sigma}^2_{v(0)} = 0.056\ (0.028)$
$\hat{\sigma}^2_{v(1)} = 0.0035\ (0.0020)$
$\hat{\sigma}_{v(0)v(1)} = 0.0081\ (0.0063)$

At the school level, the 95% coverage limits for the linear growth rates are 0.88 and 1.12, a much narrower interval than the 0.6 to 1.4 found at the pupil level. The correlation between level and growth rate at the school level is 0.58, again smaller than the corresponding correlation at the pupil level.

Let us now look at the higher level residuals (we consider the level one residuals in a slightly different way in Section 4.5). Are there any outliers and do the residuals appear to be Normally distributed? The histograms of the three pupil residuals (Figures 4.3–4.5) suggest there might be one or two outliers although extreme values of between 3.5 and 4 in absolute value are not so unusual in a sample of 407. We can also do Q-Q plots (see Chapter 2) and we find some departures from Normality for the intercept and the quadratic but not for the slope (Figures 4.6–4.8). Hence, the 95% coverage interval for the quadratic term given above should be treated with a little caution.

When we look at the school level residuals, we see no obvious outliers from the

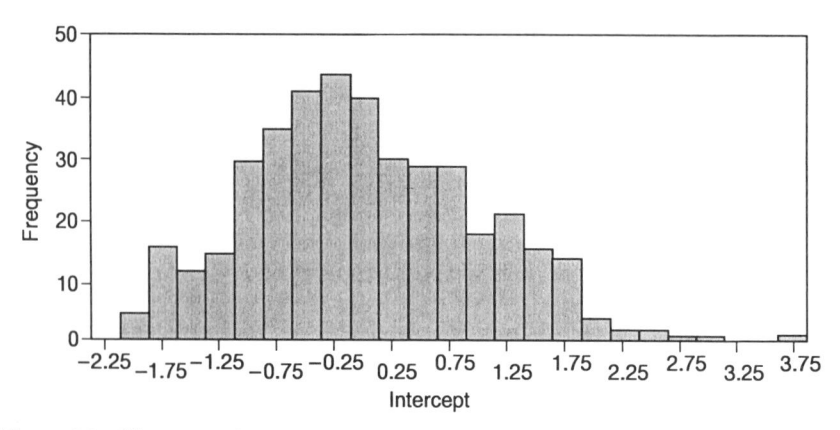

**Figure 4.3** Histogram of standardized pupil intercept residuals

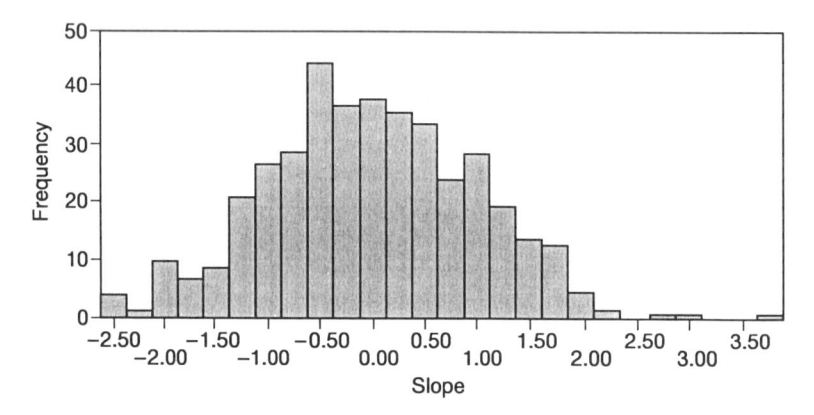

**Figure 4.4** Histogram of standardized pupil slope residuals

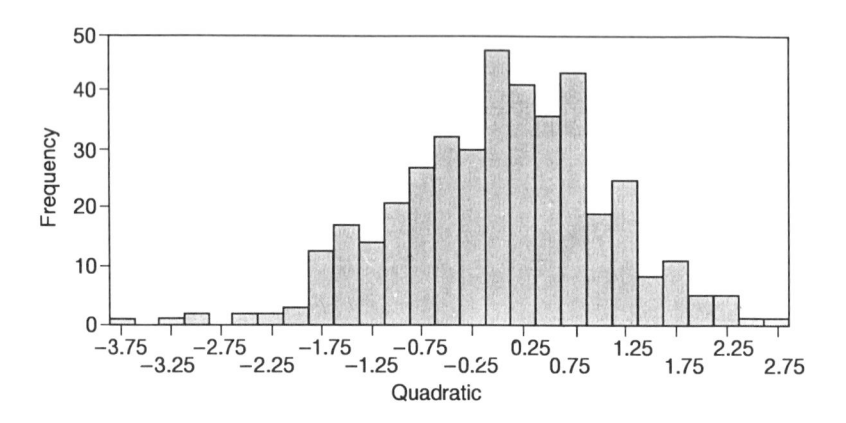

**Figure 4.5** Histogram of standardized pupil quadratic residuals

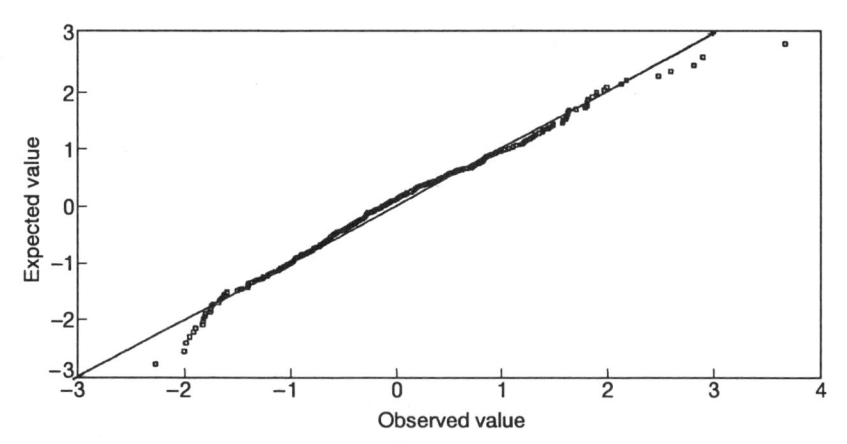

**Figure 4.6** Q–Q plot of pupil intercept residuals

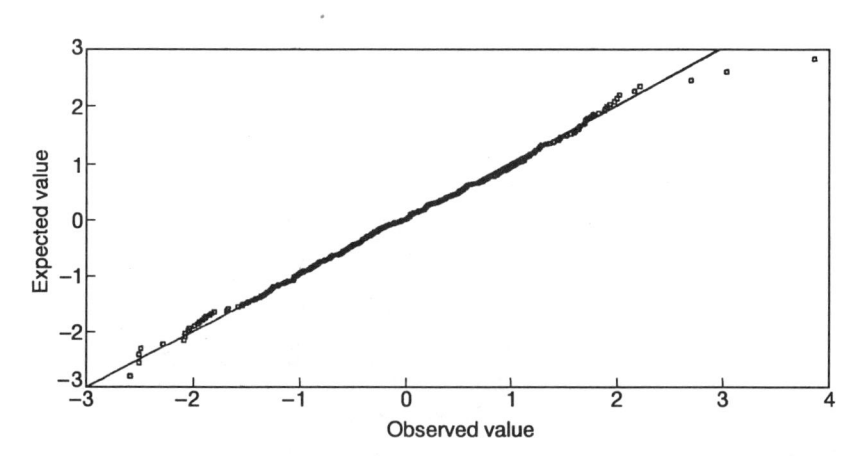

**Figure 4.7** Q–Q plot of pupil slope residuals

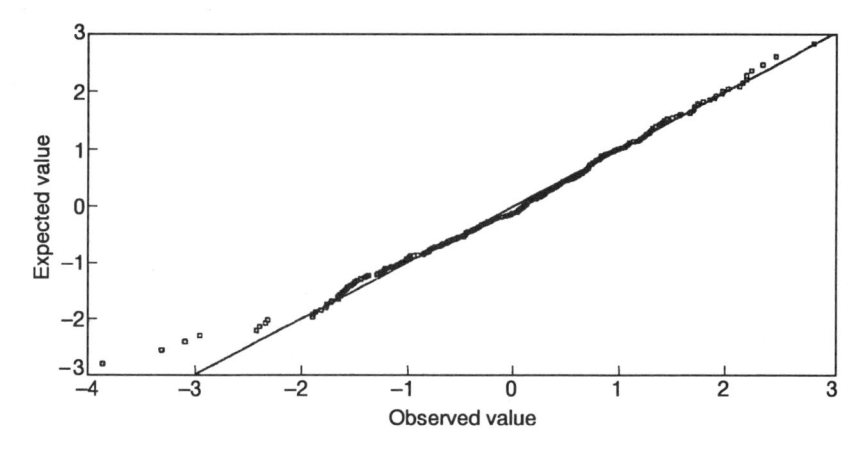

**Figure 4.8** Q–Q plot of pupil quadratic residuals

histograms (Figures 4.9 and 4.10), nor from the scatterplot of intercept against slope (Figure 4.11). Also, the Q-Q plots (Figures 4.12 and 4.13) are consistent with Normality, although, with a much smaller sample of schools than of pupils, we are likely only to pick up substantial departures from Normality.

## 4.4 Explaining variability in growth

We now know quite a lot about the structure of our data. However, we do not yet have any idea whether the variability between pupils in growth rates bears any systematic relation to what we know about the pupils – in this case, their sex and ethnic group. In other words, we have reached a position corresponding to the one we reached at the end of Section 3.5, and we now want to move on to try to account for variation at level two. We have, in fact, three explanatory variables at level two – sex, ethnic group and the interaction between them. As we saw in Section 3.6, these explanatory variables are included in the model as fixed effects

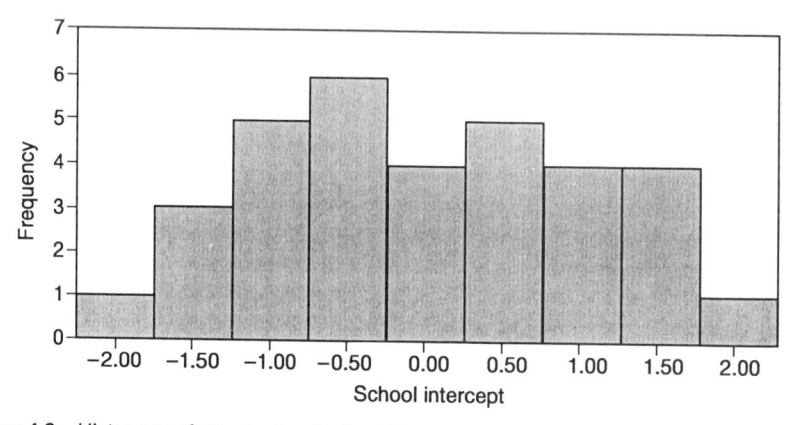

**Figure 4.9** Histogram of standardized school intercept residuals

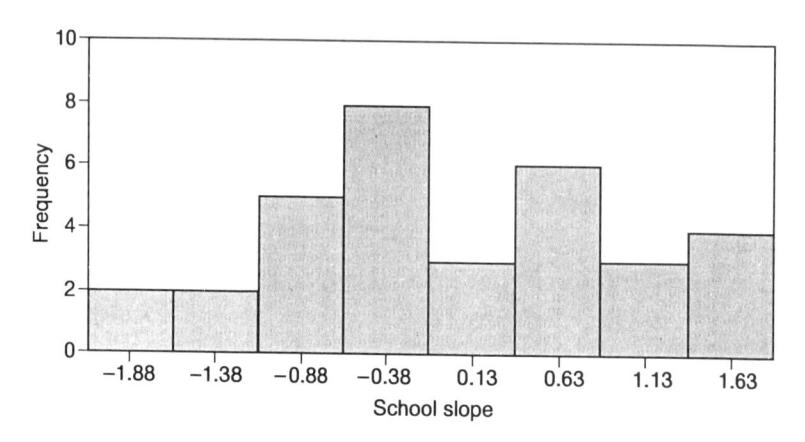

**Figure 4.10** Histogram of standardized school slope residuals

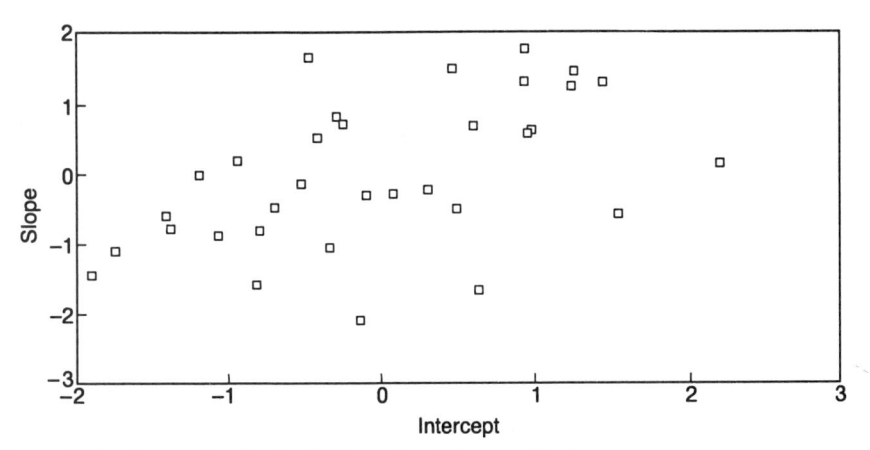

**Figure 4.11** Scatterplot of school intercept and slope residuals

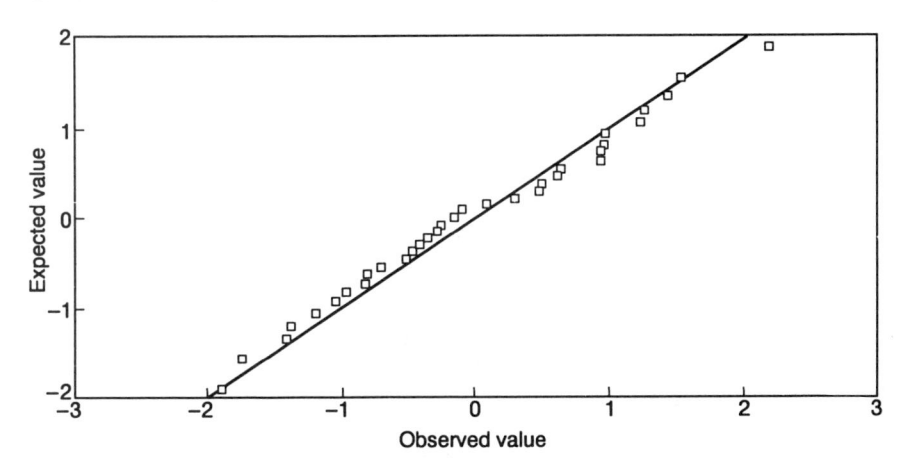

**Figure 4.12** Q-Q plot of school intercept residuals

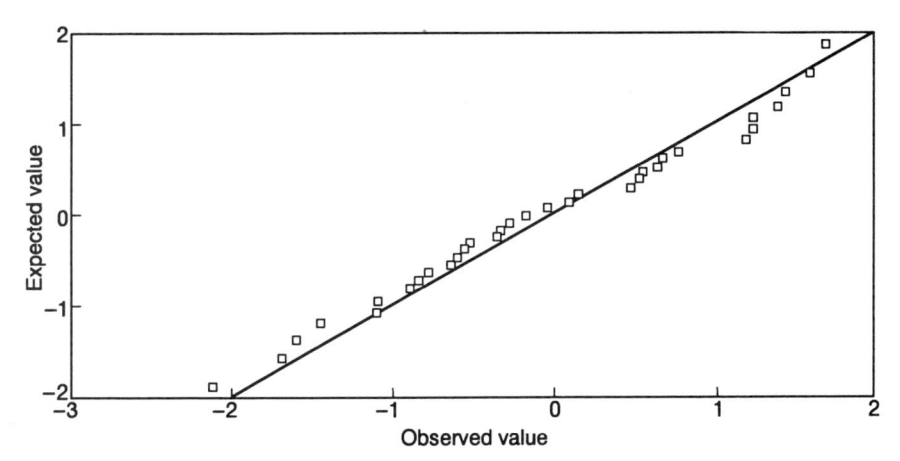

**Figure 4.13** Q-Q plot of school slope residuals

on their own (three parameters), possibly to account for variation in reading level, and as interactions with age and with age squared (six parameters altogether), possibly to account for variation in linear and quadratic reading growth. Thus, we are adding nine extra parameters and our model, including variability between schools in level but not in growth rates, now has a complicated appearance:

$$\begin{aligned}\text{READ}_{ijk} = \\ & b_{00} + b_{01}\,\text{SEX} + b_{02}\,\text{ETHNIC} + b_{03}\,\text{SEX.ETHNIC} + \\ & b_{10}\,\text{AGE} + b_{11}\,\text{AGE.SEX} + b_{12}\,\text{AGE.ETHNIC} + b_{13}\,\text{AGE.SEX.ETHNIC} + \\ & b_{20}\,\text{AGE}^2 + b_{21}\,\text{AGE}^2.\text{SEX} + b_{22}\,\text{AGE}^2.\text{ETHNIC} + b_{23}\,\text{AGE}^2.\text{SEX.ETHNIC} + \\ & v_{00k} + u_{0jk} + u_{1jk}\,\text{AGE} + u_{2jk}\,\text{AGE}^2 + e_{ijk}\end{aligned} \quad (4.5)$$

Despite its complexity, the model can be fitted easily enough. The deviance is reduced by 17.4 which, for nine degrees of freedom, is statistically significant at the 0.05 level but is, nevertheless, only a modest reduction. Our estimates are as follows (Table 4.4).

If we look at the fixed effects, we see that, apart from the constant term ($b_{00}$) and the linear term for age ($b_{10}$), the most important are $b_{13}$, $b_{23}$ and $b_{33}$ which are the coefficients for the sex ethnic group interaction and its interaction with age and age squared. All the estimated coefficients linked to sex and ethnic group separately are small compared with their standard errors. However, as we should not include an interaction in any model without also including the corresponding main effects (see p. 26), we need this rather extensive set of coefficients. We can conclude that there are different growth curves for reading for the four groups – white and black boys and girls. If we ignore all the coefficients which are smaller than their standard errors then the growth curves are:

White boys and girls: $y = 7.0 + 0.99a$ (because all the sex coefficients are small)
Black boys: $y = 7.0 + 0.94a$ (because $b_{12}$ is greater than its standard error)
Black girls: $y = 7.4 + 1.05a - 0.027a^2$ (because $b_{03}$, $b_{12}$, $b_{13}$ and $b_{23}$ are all greater than their standard errors).

In Figure 4.14, we see what these growth curves look like; reading attainment of black boys grows more slowly, and of black girls more quickly, than the other

Table 4.4  Estimates from quadratic growth curve model

| Fixed effects | | | Random effects | | |
|---|---|---|---|---|---|
| Coefficient | Estimate | s.e. | Effect | Estimate | s.e. |
| $b_{00}$ | 7.0 | 0.084 | $\sigma_v^2$ | 0.038 | 0.017 |
| $b_{01}$ | 0.098 | 0.11 | $\sigma_{u(0)}^2$ | 0.68 | 0.055 |
| $b_{02}$ | −0.084 | 0.13 | $\sigma_{u(1)}^2$ | 0.038 | 0.0041 |
| $b_{03}$ | 0.41 | 0.19 | $\sigma_{u(2)}^2$ | 0.0012 | 0.00024 |
| $b_{10}$ | 0.99 | 0.022 | $\sigma_{u(0)u(1)}$ | 0.13 | 0.013 |
| $b_{11}$ | 0.013 | 0.031 | $\sigma_{u(0)u(2)}$ | −0.01 | 0.0029 |
| $b_{12}$ | −0.043 | 0.036 | $\sigma_{u(1)u(2)}$ | −0.0016 | 0.00073 |
| $b_{13}$ | 0.10 | 0.051 | $\sigma_e^2$ | 0.13 | 0.0066 |
| $b_{20}$ | 0.0021 | 0.0055 | | | |
| $b_{21}$ | 0.0017 | 0.0080 | | | |
| $b_{22}$ | 0.0084 | 0.0091 | | | |
| $b_{23}$ | −0.027 | 0.013 | | | |

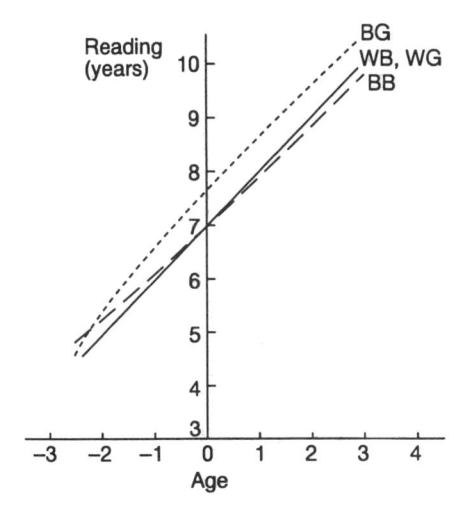

**Figure 4.14** Estimated growth curves of the four sex ethnic group combinations

groups. However, the advantage that black girls have becomes less marked as they get older (because their quadratic term is negative).

Although we have been able to account for some of the variation in reading attainment and growth, much of this variation remains unexplained. Sex, ethnic group and their interaction account for just 4% of the variation between pupils in level and in linear growth, and 6% of the variation in quadratic growth. Also, we have not attempted to explain any of the variation at the school level. One way in which this might be possible is to construct a contextual variable such as the proportion of boys in the school, but this is not explored here.

### 4.5 Complex variation at level one

In all the multilevel models we have considered in both the previous chapter and in this one, we have assumed that that there is just a constant variance term at level one, $\sigma_e^2$. Such a strong assumption is not always justified (as indicated on p. 44 in Chapter 3) and there are ways of relaxing it. We have seen that by making the coefficients of level one variables random at level two (for example, ZCURRIC in Section 3.5 and AGE in Section 4.3), we make the variation at level two vary according to the value of that level one variable. If we make the coefficient of the level one variable random at level *one* then the level one variance is no longer constant, and we have a way of building heteroscedasticity into our model. This is a device for making the model more realistic but it should be realized that no substantive interpretation can be given to the random coefficient at level one. The issue is described in more detail in Goldstein (1995, Chapter 3), who also gives a method for allowing successive level one residuals to be correlated (known as autocorrelation), which is useful for longitudinal data when the measurement occasions are close in time.

Let us see whether the assumption of constant level one variance is plausible for the reading growth data. In other words, does the residual variation between occasions within pupils vary by age?

Figure 4.15 suggests that the level one variance might be narrowing as pupils get older, although not to any marked degree. To test this, we make age random at level one and therefore include two further random parameters at level one, a variance and a covariance. We find that this does improve the fit of the model but not substantially – the fall in the deviance is 6.21 which, for two degrees of freedom, is just statistically significant at the 0.05 level. Moreover, all the estimates are very close to those presented in Table 4.4 and so there are no changes to our substantive conclusions. The level one random parameters are estimated to be:

$\hat{\sigma}_e^2 = 0.13$ (0.0085) – the level one variance for the constant
$\hat{\sigma}_f^2 = 0.00099$ (0.0039) – the level one variance for age
$\hat{\sigma}_{ef} = -0.0063$ (0.0027) – the level one covariance.

Clearly, the level one variance for age is essentially zero and can be ignored. Therefore, the level one variance is:

0.13 − 0.013 AGE (constant variance plus twice the covariance × age).

There are other ways in which we can increase the complexity of the variance structure at each level. We can, for example, allow the between pupil (or level two) variance to be different for boys and girls, or for black and white pupils. But there are always limits to how far we can go along this road, limits imposed both by the amount of data and by the wish to have a parsimonious model.

To sum up, we have shown that reading develops in a non-linear way with age, that some of the variation between pupils in level and growth is systematic in the sense that it can be explained by sex and ethnic group, that there is variation between schools in level and, possibly, in growth rates too, and there is

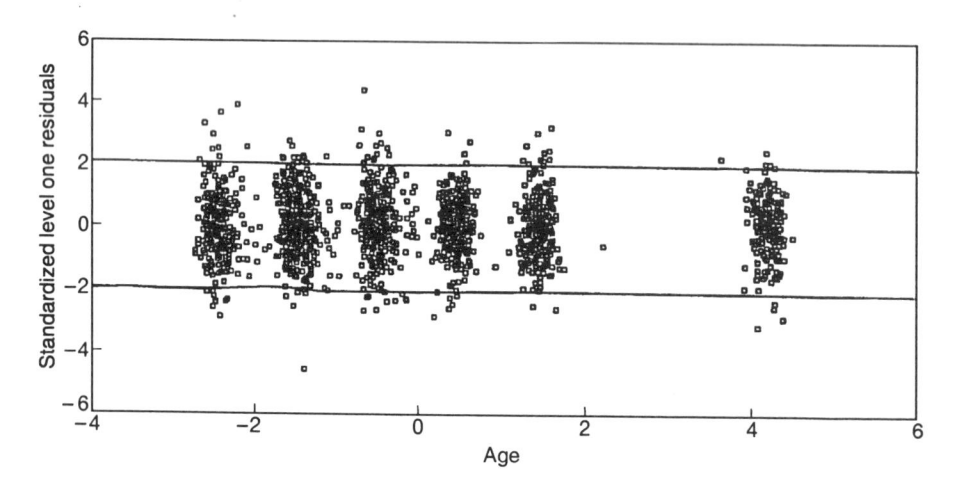

**Figure 4.15** Scattterplot of standardized level one residuals by age

## 4.6 Missing data

some evidence that the level one residuals are heteroscedastic. The approach described here in terms of growth in educational attainment could easily be adapted to deal with other repeated measures, and to contexts for which variation with time is more relevant than variation with age. For example, we might want to study the way in which schools' spending on books and equipment changes over time, and to relate this change to the funding arrangements experienced by the school.

### 4.6 Missing data

It is typical of longitudinal and repeated measures studies that subjects are lost from them over time. Sometimes this sample attrition is cumulative in that once subjects are lost from the study they never return. This is called a *monotone* pattern of missing data. But it is not unusual for subjects to be missed at one occasion, only for them to reappear in the study later, as in the example used in this chapter. Researchers, when faced with a dataset that is incomplete in some way, need to find out as much as they can about the characteristics of the data that are missing, and then to find an appropriate method of analysis for the available data. Although longitudinal studies tend to suffer more from missing data than do cross-sectional studies, they also offer more ways of finding out why some data are missing.

We can divide missing data into three classes: missing *completely at random*, missing *at random*, which together make up *ignorable* missingness, and *non-ignorably* missing. When analysing longitudinal data, missing completely at random means that those units lost from the study by any occasion $t$ are just a random subset of the original data. This is the most convenient situation but probably the least common.

Missing at random means that those units lost from the study by occasion $t$ are no different from those remaining in the study at $t$, *conditional* on their scores at earlier occasions $t-1$, $t-2$ etc. In other words, any differences between the drop-outs and those remaining can be accounted for by observable differences between these two groups earlier in the study.

If data are missing non-ignorably then this means that the growth paths of the drop-outs, if observed, would have been different from those units remaining in the study, regardless of whether they were the same or different before they were lost from the study. Non-ignorable missing data are very hard to deal with. Unfortunately, we can never be sure that the missingness is ignorable, but there are some checks we can carry out which might help us to feel more confident that such an assumption is a reasonable one.

For the reading growth data, we can draw up a two-way table of means and standard deviations, the rows of the table referring to the number of occasions for which each pupil was measured and the columns to the actual measurement occasion. We hope that, as we read down each column, the means and standard deviations show little variation from row to row within a column because that would be consistent with the data being missing completely at random. The evidence in Table 4.5 suggests that this is not a vain hope because there are no consistent differences between rows as we move from column to column, either for the means or for the standard deviations. It is important to consider

## 70 Growing and Changing

**Table 4.5** Patterns of reading scores: means and SDs

|  |  | Measurement occasion |  |  |  |  |  |
|---|---|---|---|---|---|---|---|
|  |  | 1 | 2 | 3 | 4 | 5 | 6 |
| Number of occasions measured | 1 | 4.75 (0.58) | 5.62 (0.87) | * | * | * | * |
|  | 2 | 4.74 (0.57) | 5.82 (0.75) | 6.82 (1.1) | * | * | * |
|  | 3 | 4.76 (0.55) | 5.76 (0.63) | 6.62 (0.77) | 7.52 (1.19) | * | * |
|  | 4 | 4.90 (0.84) | 5.57 (0.81) | 6.69 (0.96) | 7.50 (1.08) | 8.39 (1.22) | * |
|  | 5 | 4.74 (0.69) | 5.67 (0.76) | 6.64 (0.88) | 7.65 (1.00) | 8.65 (1.15) | 11.3 (1.50) |
|  | 6 | 4.66 (0.54) | 5.70 (0.73) | 6.67 (0.80) | 7.64 (0.92) | 8.62 (1.07) | 11.4 (1.43) |
|  | Total | 4.72 (0.61) | 5.69 (0.74) | 6.67 (0.87) | 7.62 (1.00) | 8.61 (1.12) | 11.3 (1.47) |

variation as well as level when thinking about missing data.

Another approach is to look at means and variation at occasion $t-1$ for those measured and not measured at occasion $t$ and, because we do not have monotone missing data here, the same statistics at occasion $t$ for those measured and not measured at occasion $t-1$. Some results along these lines are given in Table 4.6 and reinforce those in Table 4.5. If there had been differences between the pairs in Table 4.6, we could have split the distribution at occasion $t-1$ into categories, and then looked at differences within those categories for those measured and not measured at occasion $t$. If there were no differences within the categories of the distribution at $t-1$, then missing at random would be a reasonable assumption.

A third check is to construct an indicator variable to reflect missingness and to include this indicator variable in a multilevel model. For example, with these data, we can include a level two indicator variable ($M_1$) taking the value one if a pupil joined the sample at occasion two and zero if they joined at occasion one. Our two-level model is:

$$\text{READ}_{ij} = b_{0j} + b_{1j} \text{AGE}_{ij} + b_{2j} \text{AGE}^2_{ij} + e_{ij} \quad (4.6)$$

$$b_{0j} = b_{00} + b_{01} M_1 + u_{0j} \quad (4.7)$$

$$b_{1j} = b_{10} + b_{11} M_1 + u_{1j} \quad (4.8)$$

$$b_{2j} = b_{20} + b_{21} M_1 + u_{2j} \quad (4.9)$$

**Table 4.6** Patterns of missing reading scores: means and SDs

| Combination | Mean | SD |
|---|---|---|
| Occasion 1, not 2 | 4.71 | 0.56 |
| Occasions 1 and 2 | 4.72 | 0.62 |
| Occasion 2, not 1 | 5.65 | 0.77 |
| Occasions 1 and 2 | 5.71 | 0.72 |
| Occasion 2, not 3 | 5.62 | 0.79 |
| Occasions 2 and 3 | 5.69 | 0.74 |

## 4.6 Missing data

We are interested in $b_{01}$, which tells us if there is a mean difference between the two samples, and the coefficients of the interaction of $M_1$ with age and age squared, which tell us if the two samples grow at different rates. Not surprisingly in the light of the above results, all these coefficients were small.

Finally, we can look at patterns of correlations and covariances, taking triplets of successive occasions and calculating these statistics for the first two occasions of each triplet, depending on whether or not the pupil was tested on the third occasion of the triplet, as shown in Table 4.7.

**Table 4.7** Patterns of correlations and covariances

| Combination | Correlation (t−2,t−1) | Covariance |
| --- | --- | --- |
| Occasions 1 and 2, not 3 | 0.80 | 0.37 |
| Occasions 1, 2 and 3 | 0.56 | 0.25 |
| Occasions 4 and 5, not 6 | 0.90 | 1.10 |
| Occasions 4, 5 and 6 | 0.89 | 0.91 |

Table 4.7 does reveal some differences at occasions one and two, although the first row of the table is based on just 11 pupils and so these differences could have arisen by chance.

Most of what we know about the missingness in these data suggests that it is ignorable. Hence, any biases in the analyses presented in previous sections are probably small, and the analyses are as efficient as possible. We can compare the results obtained from the available data with those obtained from the complete data, based on those 126 pupils with data at each occasion. The comparison is given in Table 4.8 in terms of the ratios of the standard errors for the fixed and random effects for the complete data compared with the available data. All these ratios are greater than one, sometimes substantially so, which shows how much more efficient the analysis using all the available data is.

**Table 4.8** Ratios of standard errors for complete and available data

| Fixed effects | | Random effects | |
| --- | --- | --- | --- |
| Coefficient | Ratio | Effect | Ratio |
| $b_{00}$ | 1.48 | $\sigma_v^2$ | 1.73 |
| $b_{01}$ | 1.63 | $\sigma_{u(0)}^2$ | 1.35 |
| $b_{02}$ | 1.62 | $\sigma_{u(1)}^2$ | 1.41 |
| $b_{03}$ | 1.53 | $\sigma_{u(2)}^2$ | 1.17 |
| $b_{10}$ | 1.50 | $\sigma_{u(0)u(1)}$ | 1.38 |
| $b_{11}$ | 1.81 | $\sigma_{u(0)u(2)}$ | 1.10 |
| $b_{12}$ | 1.39 | $\sigma_{u(1)u(2)}$ | 1.32 |
| $b_{13}$ | 1.55 | $\sigma_e^2$ | 1.52 |
| $b_{20}$ | 1.29 | | |
| $b_{21}$ | 1.53 | | |
| $b_{22}$ | 1.18 | | |
| $b_{23}$ | 1.31 | | |

If, as a result of the kinds of investigations of missingness described here, we do not feel confident about the assumption of ignorability, then we have to find other approaches. If the proportion of missing data is small then the bias introduced by assuming ignorability will usually be small. However, in situations of

72  *Growing and Changing*

considerable missing data, then the methods described by Little and Schenker (1995) may be needed, which incorporate some strong assumptions.

## 4.7 Interpretational issues

The validity of the results in this chapter rests on two assumptions. First, we assume that the results are robust to changes in the arbitrary scale for the response. Second, we assume that the statistical model used – an unconditional, age-related model – is appropriate.

We can make some progress testing the first assumption by trying different scales for the response. Rather than a scale with a constant coefficient of variation (c.v.), we can try scales with an increasing and a decreasing coefficient of variation, but at the same time maintaining the condition that the variance of the scale increases with age. Thus, we see what happens when we allow the c.v. to increase from 0.13 to 0.18 over the age range, and when we allow it to decrease from 0.18 to 0.13. We fit the same quadratic model as before, but we find that some of our results vary according to the scale adopted. For example, when we consider the relative amounts of reading growth made by the four sex ethnic groups, we find some differences. If we substitute for age in our final model, using the fixed effect estimates in Table 4.4 for the scale with a constant c.v., and the corresponding estimates for the other two scales, we get predicted values for each group at particular ages. So, from the growth curves given on p. 66, our predicted values at the beginning of school, when age on our centered scale is –2.41, are:

White boys and girls: $y = 7.0 + 0.99 (–2.41) = 4.61$
Black boys: $y = 7.0 + 0.94 (–2.41) = 4.73$
Black girls: $y = 7.4 + 1.05 (–2.41) – 0.027 (2.41 \times 2.41) = 4.71$

We can go through the same process for age at the end of infant school and at the end of junior school. This enables us to calculate relative amounts of growth over these two periods of schooling. In fact, for each of the three scales, the white boys and girls make equal amounts of growth. However, the results for the black boys and girls are different, as shown in Table 4.9. For the infant school period, the differences are small and unimportant. But for junior school, we see that black boys fall behind white pupils for two of the scales, whereas they make slightly more growth for the scale with a decreasing c.v. Also, black boys' reading grows more than black girls' when the c.v. decreases,

Table 4.9 Changes in relative growth (reading 'years') with changes in scale

| Group | Infant school | | | Junior school | | |
|---|---|---|---|---|---|---|
| | Constant c.v. | Increasing c.v. | Decreasing c.v. | Constant c.v. | Increasing c.v. | Decreasing c.v. |
| White pupils | 0 | 0 | 0 | 0 | 0 | 0 |
| Black boys | –0.17 | –0.14 | –0.23 | –0.14 | –0.19 | 0.05 |
| Black girls | 0.32 | 0.43 | 0.34 | –0.14 | –0.06 | –0.18 |

## 4.7 Interpretational issues

the same amount when the c.v. is constant, and less when the c.v. increases.

Turning to the random part of the model, we find differences there too. There is no evidence of complex variation, or heteroscedasticity, at level one for the scales with increasing and decreasing c.v., whereas there was, as we saw in Section 4.6, when the c.v. was constant. Also, the correlations between the different effects vary substantially, particularly those involving the quadratic term, as shown in Table 4.10. This reinforces the point made earlier that we should be very wary of basing any theories about growth on these correlations.

The fact that changes of scale can affect the results from multilevel growth curve models implies that the approach is most suitable when there is a fixed scale. Unfortunately, as we have seen, fixed scales are uncommon in educational research. There are circumstances when we can reasonably suppose that criterion-referenced tests – tests based on what pupils can do rather than how they compare with other pupils – provide a fixed scale over age. However, criterion-referenced tests are not commonly used, researchers mainly relying on norm-referenced attainment tests of the kind used here. In these circumstances, researchers need to try different assumptions about their underlying scale, and to rely only on those findings that are robust to changes in those assumptions. In the example used here, the group differences in growth during infant school were more robust than those found for junior school.

However, there is an alternative way of approaching these types of questions which is to use the so-called conditional approach of the kind described in Chapters 2 and 3. In other words, we look at attainment at occasion $t$ for fixed values of attainment at occasion $t-1$ to define progress, and then see whether our explanatory variables – ethnic group and sex here – are related to progress. Our multilevel structure here is generated by the location of pupils within schools, but not by the repeated measurement of the pupils. At its simplest, with a random intercept but no random slopes, the model is:

$$\text{READ}_{ij}^{(t)} = b_{0j} + b_1 \text{READ}_{ij}^{(t-1)} + b_2 \text{SEX}_{ij} + b_3 \text{EG}_{ij} + b_4 \text{SEX.EG}_{ij} + e_{ij} \qquad (4.10)$$

The estimates of $b_2$, $b_3$ and $b_4$ are of most interest here.

Although both the growth curve and the conditional approaches are used to tackle the same underlying issue – what are the correlates of growth? – they are, nevertheless, different. The strengths of multilevel growth curve models lie in their ability to exploit all the available data and to extend our knowledge about the patterns of growth in attainments. One of the strengths of the conditional approach is to make the absence of a fixed scale over time less important because

**Table 4.10** Changes in correlations with changes in scale

| Correlation | Level two (pupil) | | | Level three (school) | | |
|---|---|---|---|---|---|---|
| | Constant c.v. | Increasing c.v. | Decreasing c.v. | Constant c.v. | Increasing c.v. | Decreasing c.v. |
| Level and linear | 0.82 | 0.85 | 0.77 | 0.58 | 0.78 | 0.23 |
| Level and quadratic | −0.37 | 0.20 | −0.64 | n.a. | n.a. | n.a. |
| Linear and quadratic | −0.25 | −0.05 | −0.67 | n.a. | n.a. | n.a. |

## 74 Growing and Changing

linear transformations of the response in a regression model have no effect on the conclusions. Also, the conditional approach is better suited to causal modelling because, by conditioning on earlier measures (including the 'pre-test'), we get closer to the notion of randomization in experimental research. The advantages and disadvantages of these two approaches for the measurement and explanation of growth are discussed in more detail in Plewis (1996b).

This brings us to the end of our detailed introduction to multilevel modelling for educational data. However, in Chapter 8, we meet less detailed descriptions of other ways in which multilevel models can be applied to rather more complicated data types and structures. But, for now, we move away from methods for continuous responses to methods and models for categorical data.

### Exercises

The answer to the exercise marked '#' can be found at the end of the book. Exercises 2 and 3 require access to a specialist multilevel modelling package or procedure.

**1** When mathematics is used as the response for the study described in Chapter 4, the following table, similar to Table 4.4, is obtained. Give a brief interpretation of this table.

| Fixed effects | | | Random effects | | |
|---|---|---|---|---|---|
| Coefficient | Estimate | s.e. | Effect | Estimate | s.e. |
| $b_{00}$ | 7.3 | 0.090 | $\sigma^2_{v(0)}$ | 0.074 | 0.032 |
| | | | $\sigma^2_{v(1)}$ | 0.0038 | 0.0018 |
| | | | $\sigma_{v(0)v(1)}$ | 0.016 | 0.0071 |
| $b_{01}$ | −0.25 | 0.11 | $\sigma^2_{u(0)}$ | 0.64 | 0.054 |
| $b_{02}$ | −0.41 | 0.13 | $\sigma^2_{u(1)}$ | 0.018 | 0.0031 |
| $b_{03}$ | 0.42 | 0.18 | $\sigma^2_{u(2)}$ | 0.00028 | 0.00033 |
| $b_{10}$ | 1.06 | 0.022 | $\sigma_{u(0)u(1)}$ | 0.10 | 0.011 |
| $b_{11}$ | −0.088 | 0.031 | $\sigma_{u(0)u(2)}$ | −0.0051 | 0.0028 |
| $b_{12}$ | −0.12 | 0.027 | $\sigma_{u(1)u(2)}$ | −0.000045 | 0.00063 |
| $b_{13}$ | 0.12 | 0.044 | $\sigma^2_{e(0)}$ | 0.18 | 0.011 |
| | | | $\sigma^2_{e(1)}$ | 0.0081 | 0.0048 |
| | | | $\sigma_{e(0)e(1)}$ | 0.012 | 0.0039 |
| $b_{20}$ | −0.015 | 0.0054 | | | |
| $b_{21}$ | 0.018 | 0.0091 | | | |
| $b_{22}$ | 0.022 | 0.0080 | | | |
| $b_{23}$ | −0.028 | 0.013 | | | |

This table is based on 1757 observations, one outlier having been omitted. The fall in the deviance from making age random at level one is 12.5 (2 df), and the fall in the deviance from making age random at level three is 11.9 (2 df). See Plewis (1996b) for further details.
(#)

**2** Using dataset 4.1 on the disk, transform the reading attainment variable to

Normality at each occasion and see whether this makes the level two residuals closer to Normality in the three-level model of p. 61.

3   Investigate how the model of p. 66 changes when reading attainment is standardized to have mean zero and variance one at each occasion (i.e use dataset 4.1 with ZREAD as response).

4   A researcher interested in the way attitudes to science change as pupils move through secondary school, and the way in which these attitudes vary at the school level over time, uses the design set out as Figure 4.2. The level one variable is pupil age, the level two variable is pupil sex, the level three variable is time and the level four variable is school type (selective and non-selective). The researcher hypothesizes:
   (a) that attitudes change linearly with pupil age, becoming more positive as pupils move through secondary school, and this linear trend is more marked for boys than it is for girls;
   (b) that attitudes change linearly with time, becoming less positive;
   (c) that attitudes are more positive for boys than for girls, and this difference is greater in selective schools;
   (d) that attitudes are more positive in selective schools.

Write down a multilevel model corresponding to these hypotheses, and indicate what signs you expect the coefficients of the model to have.

# 5
# Two by Two Tables and Beyond: Modelling Binary Responses

## 5.1 Introduction

Up to now, we have been concerned with the analysis of studies where the response is measured on a continuous scale (although there have been no restrictions on the scale for the explanatory variables). However, in common with most social research, many of the variables we meet in educational research have just a few, often only two, categories. This chapter covers some basic descriptive methods for the analysis of binary responses when there is just one categorical, but not necessarily binary, explanatory variable. It also shows how variation in a binary response can be modelled by linking the response to one or more explanatory variables. Often, the best way of making this link is to use the method of *logistic regression*.

The chapter starts by looking in some detail at two-way tables. It then goes on to show how to model these tables, and how to extend the ideas to *multi-way* tables. Tables for dependent responses, often generated from longitudinal data, are then introduced. The chapter ends by drawing out the distinction between *association* and *agreement* for binary variables.

## 5.2 The two by two table

On their own, we represent categorical variables in terms of counts or frequencies, and summarize them using proportions, probabilities or percentages. In combination, we represent the relationships between them in terms of tables of counts, which are often known as *contingency tables*. A good introduction to the architecture of contingency tables is given by Marsh (1988). There is a wealth of information in even the simplest two-way table, a two by two table connecting a binary response to a binary explanatory variable. This descriptive information, usually expressed in terms of the *marginal* and *conditional* probabilities, should never be ignored. Too often, researchers, when confronted by a contingency table, immediately launch into a test of independence of some kind without considering what else they could learn from the table.

Table 5.1 is a cross-tabulation of two binary variables. The response is whether or not a child is *perceived* by their teacher to have a behaviour problem, and the explanatory variable is ethnic group (white and black, defined in the same way as in Chapter 4). The data were obtained for 172 boys in reception classes and come, like many of the examples in this chapter, from the Tizard *et al.* (1988) study described on p. 57. We see from the marginal percentages that the majority of boys are not considered to have a behaviour problem (70%) and that 63% of this sample are white. From the cell sizes, we can easily see that more than half the black boys, but only a relatively small proportion of white boys, are thought by their teachers to have behaviour problems. The conditional probabilities, which always sum to one within each category of the explanatory variable, are shown in square brackets. It does appear that the two variables are related because there is a marked difference in the two sets of conditional probabilities. We would like a convenient way of summarizing the strength of this association.

One way of using the information in contingency tables, which is useful, not only for two by two tables but also for more complicated ones, is to calculate *odds* and *relative odds*. In the above table, the odds of a white boy being seen to have a behaviour problem are 19/90 = 0.21 (or 0.21 to 1 which, in betting terms, is about 5 to 1 against or much less than even money). The corresponding odds for black boys are 33/30 = 1.1 (1.1:1, equivalent to 11 to 10 on, or a little better than even money). Note that odds are not probabilities; they are not restricted to the range zero to one. The relative odds of a black boy compared with a white boy being seen as having a behaviour problem are 1.1/0.21 or 5.2:1. In other words, a black boy is 5.2 times more likely than a white boy to be seen as having a behaviour problem. Equally, boys with behaviour problems are 5.2 times more likely to be black rather than white, compared with boys without behaviour problems. Relative odds are symmetrical in that sense; like the Pearson correlation of Chapter 2, they do not vary according to which variable is the response and which is explanatory.

More formally, if we label the cells from left to right in Table 5.1 as $a$ (=90) and $b$ (=30) in the first row, and as $c$ (=19) and $d$ (=33) in the second row, then the relative odds, $\alpha$, sometimes known as the *cross-product ratio* and sometimes as the *odds ratio*, can be written as:

$$\alpha = ad/bc \tag{5.1}$$

The minimum value of $\alpha$ is zero, attained when either cell $a$ or cell $d$ is zero, the maximum value is infinity, attained when either cell $b$ or cell $c$ is zero. A value of

**Table 5.1** Behaviour problems by ethnic group: cell counts, conditional probabilities and marginal percentages

| Behaviour problems | Ethnic group | | Total |
|---|---|---|---|
| | White | Black | |
| No | 90 [0.83] | 30 [0.48] | 120 (70%) |
| Yes | 19 [0.17] | 33 [0.52] | 52 (30%) |
| Total | 109 (63%) | 63 (37%) | 172 (100%) |

## 5.2 The two by two table

one for the relative odds, α, is equivalent to saying that there is no association between the two binary variables, because the odds are the same for each category of the explanatory variable.

To have a measure of association which can vary from zero to infinity and thus cannot be negative is not ideal. We can make it symmetrical about zero by taking logs, although the log of zero is not defined. More usefully, we can use another coefficient, Yule's $Q$, which is defined as:

$$Q = (ad - bc)/(ad + bc) \tag{5.2}$$

$$= (\alpha - 1)/(\alpha + 1) \tag{5.3}$$

$Q$ is $-1$ when $\alpha$ is zero and $+1$ when $\alpha$ is infinity. When $Q$ is zero (and $\alpha = 1$), there is no association. For Table 5.1, $\hat{Q}$ is 0.68, which suggests that ethnic group is a good predictor of perceived behaviour problems in this sample. (Later in this chapter we have cause to revise this interpretation.)

The data in Table 5.1 are, of course, sample data. Thus, the values of α and $Q$ are estimates with sampling errors. We can write their standard errors as follows:

$$\text{s.e. } (\hat{\alpha}) \cong \hat{\alpha} \sqrt{[1/a + 1/b + 1/c + 1/d]} \tag{5.4}$$

and

$$\text{s.e. } (\hat{Q}) \cong 1/2(1 - \hat{Q}^2) \sqrt{[1/a + 1/b + 1/c + 1/d]} \tag{5.5}$$

providing the samples are large, say greater than 100, and providing none of the cells are zero.

Thus, for Table 5.1, the standard errors are 1.9 for $\hat{\alpha}$ and 0.10 for $\hat{Q}$. For large samples, we can use the usual Normal approximation to construct a confidence interval for $Q$. Thus, a 95% confidence interval for $Q$ in Table 5.1 is about 0.48 to 0.88. This, in turn, gives an approximate 95% confidence interval for α of 2.9 to 15.7. Notice that the confidence interval for α is not symmetrical about $\hat{\alpha}$ (=5.2) and it does not include one.

For many tables such as Table 5.1, it is clear that the two variables are associated, and the most interesting issue is how strong is the association. (Just as in regression, the size of the regression coefficient is usually more relevant than whether the coefficient is different from zero.) Nevertheless, it is sometimes useful to do a test of the hypothesis of no association, or of independence between the two variables, and the method for doing this is given below. In essence, what such a test is doing is to test whether α differs from one, or whether $Q$ differs from zero. (For these data, our 95% confidence intervals for $Q$ and α lead us to reject these hypotheses at the 5% level.)

If the two binary variables are independent, then the *joint* probability of a 'yes' response for both the variables is equal to the product of the two corresponding marginal probabilities. In other words, the expected cell counts, given independence, are equal to the product of the two marginal counts divided by the sample size. (Always use the cell counts in the calculations, never the cell percentages or probabilities.) So for Table 5.1, the expected cell sizes are (to the nearest whole number) 76 for $a$ (i.e. $120 \times 109/172$), 44 for $b$, 33 for $c$ and 19 for $d$. If we call the observed cell counts $o_{ij}$ and the expected counts $e_{ij}$ ($i$, the row indicator, $= 1,2$; $j$, the column indicator, $= 1,2$) then we calculate:

$$X^2 = \Sigma (o_{ij} - e_{ij})^2 / e_{ij} \tag{5.6}$$

and $X^2$ has a $\chi^2$ distribution with one degree of freedom. (There is one degree of freedom because we can 'choose' the count for just one cell in a two by two table, once we have fixed the margins.) Consequently, we compare the value of $X^2$ with the chosen critical value of the $\chi^2$ distribution, which is 3.84 if we work with a significance level of 0.05. The contributions of each of the cells to the total value of $X^2$ (i.e. $(o_{11} - e_{11})^2/e_{11}$ etc) can sometimes help us to understand better where the departure from independence comes from. These contributions are shown in Table 5.2 for the observed values in Table 5.1. We see that the largest contribution comes from the 'yes/black' cell. The total value – 23.1 – is, of course, highly statistically significant. These data do not support the null hypothesis of independence of perceived behaviour problems and ethnic group.

**Table 5.2** Contributions to $X^2$ from Table 5.1

| Behaviour problems | Ethnic group |       | Total |
|---|---|---|---|
|  | White | Black |  |
| No  | 2.56 | 4.43 |  |
| Yes | 5.91 | 10.2 |  |
| Total |  |  | 23.1 |

It is very important to understand that $X^2$ is not a measure of association, merely a statistic that can be used to infer whether the association between two variables is greater than zero in the population. The hypothesis of independence, or no association, is often not especially interesting, partly because it is increasingly likely to be rejected as the sample size increases. If we multiplied all the observed cell counts by ten in Table 5.1, the value of $X^2$ would also be multiplied by ten, but the estimates of $\alpha$, and hence of $Q$, would be unchanged (although their standard errors would be smaller). We should also note that when expected cell sizes are small, say less than five, some textbooks and computer manuals recommend using a correction to $X^2$ called Yates' correction. However, modern statistical theory indicates that Yates' correction is not especially helpful. On the other hand, if the total sample size is small, then Fisher's exact test is more appropriate than the usual $\chi^2$ test of independence and this is described in Blalock (1979).

## 5.3 Larger contingency tables

Let us now move on to $2 \times c$ contingency tables, where $c$ is the number of columns, and where $c$ is greater than two. (We could also call these tables $r \times 2$ tables where $r, r > 2$, is the number of rows.) Rarely is $c$ greater than five. Let us consider first the case where the variable defining the columns is unordered. We would have obtained such a table if we had had more than two ethnic groups in the previous section. Some measures of association have been proposed for this situation (see Blalock, 1979) but none are entirely satisfactory. This is partly because there are now $c-1$ degrees of freedom. This in turn means that there are $c-1$ independent relative odds for the table, and so we cannot expect that one coefficient will summarize the data adequately. We can estimate these relative odds, and we can obtain an overall value of $X^2$, together with contributions to $X^2$ from each cell, using the same methods that we used in Section 5.2. Now we compare $X^2$ with the critical value for the $\chi^2$ distribution with $c-1$ degrees of freedom.

There is more scope for analysis if the explanatory, or column, variable is ordered. Consider Table 5.3: this gives the numbers and prevalences of aggressive and distractable behaviour problems for the first three years of schooling. (We treat these data as if they come from a repeated cross-sectional design and ignore, for now, the fact that some of the pupils were rated on more than one occasion.) There appears to be an increase from year to year. How large is the association with school year, and is the increase in prevalence more than we would expect by chance?

**Table 5.3** Aggressive and distractable behaviour problems by school year: number and prevalence

| Behaviour problems | School year | | |
|---|---|---|---|
| | Reception | Year one | Year two |
| Yes | 49 [0.14] | 52 [0.16] | 54 [0.21] |
| Total | 354 | 327 | 259 |

A number of coefficients of association have been proposed for tables of this kind and most are produced routinely by statistical packages. Kendall's tau-*c* ($\tau_c$), a kind of rank order correlation, is as good as any. The calculation of $\tau_c$ is a little long-winded without a computer and there is little to be gained by presenting the formula. The value of $\tau_c$ for Table 5.3 is 0.059 with a standard error of 0.027, indicating only a little association or trend across the three years.

Clearly, the association in Table 5.3 is small but it may be more than we would expect by chance. We can use a Mantel–Haenszel test here, which has a $\chi^2$ distribution with one degree of freedom. It is based on the assumption that scores can be assigned to the column variable, and the default scores in computer packages are the integers from one up to *c* (three in our case). The Mantel–Haenszel test is essentially a chi-square test for trend in two by *c* tables, and is described in more detail in Armitage and Berry (1987). For Table 5.3, the Mantel–Haenszel chi-square is 5.13, giving a *p*-value of 0.03. The usual chi-square test for independence gives a value of 5.46 with two degrees of freedom and a *p*-value of 0.07. Hence, the Mantel–Haenszel test, by exploiting the order in school year, is more powerful here. If we had been able to exploit the longitudinal aspects of the data, then we would expect a lower *p*-value because longitudinal designs are more efficient than repeated cross-sections for analysing trends of this kind. Both Kendall's $\tau_c$ and the Mantel–Haenszel test are available within the SPSS CROSSTABS procedure.

## 5.4 Introducing logistic regression

Our main concern up to this point has been to describe associations between a binary response and a single categorical explanatory variable. As useful as this approach can be, we know from Chapter 2 that it is a rather restrictive way of analysing data. Ideally, we would like a technique that we can apply to binary responses akin to the simple and, especially, the multiple regression methods we described for continuous responses. In other words, how can we model binary responses?

We can think of the data in Table 5.1 in two ways. The first way is to regard them as two proportions – the proportion of behaviour problems in two independent samples, a sample of white boys and a sample of black boys. The second way is to think of the data as 172 observations, with the response always 0 (no problem) or 1 (problem) and the explanatory variable also either 0 (white) or 1 (black). So we want to model either the variation in the proportion of perceived behaviour problems or the variation in the probability of being perceived to have a problem, and these are the same models. However, proportions and probabilities differ from continuous responses in some important ways. They are bounded by zero and one whereas continuous variables can, in principle, vary between plus and minus infinity. Hence, we cannot assume Normality for a proportion. Instead, we must recognize that proportions have a Binomial distribution and one of the properties of the Binomial distribution is that, unlike the Normal distribution, the mean and the variance are not independent. The mean is $P$ and the variance is $P(1-P)/n$ where $n$ is the number of observations or 'trials' and $P$ is the probability of 'success' (problem, here) on any one trial (boy, here). Alternatively, we can say that a zero–one variable has a mean $P$ and variance $P(1-P)$, sometimes known as a Bernoulli distribution.

In Chapter 1, we mentioned that there are a number of ways in which we can link our response to our explanatory variables, and that there are a number of distributions to choose from for the response. With proportions as responses, we assume a Binomial distribution and we use what is known as a *logistic* or *logit* link. In other words, we transform a proportion $P$ to:

$$\log_e [P/(1-P)] \tag{5.7}$$

Note that the expression in square brackets is just the odds of a success or, in the context of Table 5.1, the odds of a behaviour problem. The logit link stretches the scale for the proportion from a scale which has a range from zero to one, to a scale which can vary from plus to minus infinity. Figure 5.1 plots the logit of $P$ against $P$ itself. We see the curvature at the extreme values of $P$ as the curve approaches plus and minus infinity, and that logit $P=0$ when $P=0.5$. We also notice that between the values for $P$ of about 0.2 and 0.8, the curve is, in fact, more or less a straight line. One implication of this is that doing ordinary regression with the observed proportion $p$ as the response will work quite well for values of $p$ in this range. Problems arise, however, for large (or small) observed values of $p$ because we can then get predicted values of $P$ outside the permitted range of 0 to 1, something which cannot happen with the logit scale.

A logistic regression, or logit, model for Table 5.1 has the following form:

$$\log [P_i/(1-P_i)] = \text{logit } P_i = a + bx_i \tag{5.8}$$

where $P_i$ is the probability of a behaviour problem for observation $i$ given this model, and $x_i$ is ethnic group. We do not transform the *observed* proportions. Thus, the parameter $a$ gives the log odds of a behaviour problem for white boys ($x=0$) and $b$ shows how much higher the log odds are for black boys ($x=1$).

We can write the model in terms of odds as:

$$P_i/(1-P_i) = \exp(a+bx_i) \tag{5.9}$$

## 5.4 Introducing logistic regression

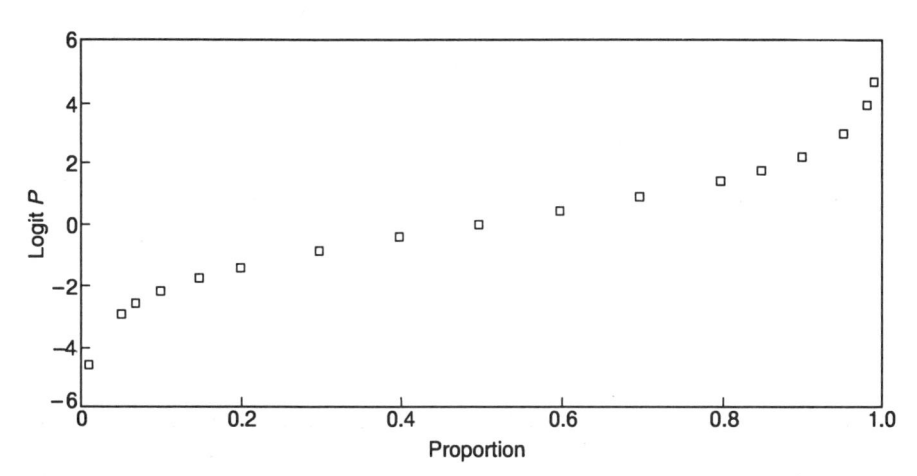

**Figure 5.1** Plot of logit $P$ against $P$

or in terms of the probability of 'success' as:

$$P_i = \exp(a + bx_i)/(1 + \exp(a + bx_i)) \quad (5.10)$$

which implies that the probability of 'failure' is:

$$1 - P_i = 1/(1 + \exp(a + bx_i)) \quad (5.11)$$

Note that we do not usually include a residual term when we write down a logistic regression model, instead expressing the model in terms of population probabilities. However, we could write

$$p_i = P_i + f_i = \exp(a + bx_i)/(1 + \exp(a + bx_i)) + f_i \quad (5.12)$$

although $f_i$ is not, of course, Normally distributed in this model.

The logistic regression model is an example of a non-linear model; it is a *generalized linear model* rather than a general linear model of the kind described in Chapter 2. This means that we need an iterative algorithm to estimate the parameters, just as we did for the multilevel models of Chapters 3 and 4. We need not concern ourselves with the details of the algorithm, nor with the theory behind it, which is based on the principle of maximum likelihood and is set out in, for example, McCullagh and Nelder (1989). Most statistical packages contain a logistic regression procedure. SPSS has two options for modelling binary responses. The first is the logit option in the LOGLINEAR menu, which is better for proportions. The second is the logistic regression procedure in the REGRESSION menu, which is better for 0–1 data. However, some care is needed when specifying the models and interpreting the coefficients from these SPSS options because the models can be parameterized in different ways. In other words, the baseline category for the explanatory variables can vary, or the effect of an explanatory variable can be measured as a deviation from a mean of zero. The same comments apply when comparing results from different statistical packages.

For Table 5.1, our fitted model, with standard errors in brackets, is:

$$\text{logit } P = -1.56 + 1.65 \, EG \qquad (5.13)$$
$$(0.25) \, (0.36)$$

which we can interpret as the log odds of a white boy ($EG=0$) seen as having a behaviour problem being equal to $-1.56$, which in turn means that the odds of a white boy having a behaviour problem are $\exp(-1.56) = 0.21$. The log odds of a black boy ($EG=1$) having a behaviour problem are 1.65 higher, i.e. 0.09, which means that the odds of a black boy having a behaviour problem are $\exp(0.09) = 1.1$. Alternatively, we can say that the odds for black boys are $\exp(1.65) = 5.21$ times as high as they are for white boys. In other words, the relative odds of a black boy being perceived by their teachers to have a problem compared with white boys are 5.21.

These odds and relative odds, estimated from the model, are identical to those estimated from Table 5.1. This is because we have just one degree of freedom in Table 5.1, and so our model is what we call a *saturated* model for which the expected and observed counts are identical. That is, after fitting a term for ethnic group, there is no residual variation. Yet another way of expressing this is to say that, before fitting the ethnic group term, the deviance (see Section 3.5) is 22.8, and afterwards, it is zero. The decline in the deviance is very close to the chi-square value from the test of independence, with one df, which we found to be 23.1.

Let us now turn to Table 5.3. Our model is now:

$$\text{logit } P_i = a + bx_i \qquad (5.14)$$

where $P_i$ is the probability of pupils who are aggressive and distractable and $x_i$ is school year, taking the values 0, 1 and 2. Note that $b$ is now the linear effect of year on the logit of $P$, which means that there is still one degree of freedom remaining after year has been fitted, and so the model is unsaturated. (We could have ignored the order and represented year by two dummy variables; this would once more have been a saturated model.) The fitted equation is:

$$\text{logit } P = -2.1 + 0.25 \, \text{YEAR} \qquad (5.15)$$
$$(0.24) \, (0.11)$$

and so the odds of a problem are multiplied each year by 1.28 ($\exp(0.25)$) starting from 0.12 ($\exp(-2.1)$) in the reception year (YEAR = 0). Thus, the predicted values are 0.12, 0.16 and 0.20, which are very close to the observed values of 0.14, 0.16 and 0.21 in Table 5.3. The decline in the deviance from the introduction of YEAR is 5.12, which is essentially the same as the value of the Mantel–Haenszel statistic obtained earlier. The residual deviance is 0.21 with one degree of freedom, which is very small and confirms that the effect of year on prevalence can be adequately represented by a linear term, rather than by two dummy variables.

## 5.5 Logistic regression with more than one explanatory variable

So far, our use of logistic regression has not told us a lot more than we were able to find out from the descriptive approaches of Sections 5.2 and 5.3. However, suppose we extend Table 5.1 to include girls as well as boys so that there are now two explanatory variables – ethnic group and sex. Our questions are familiar

## 5.5 Logistic regression with more than one explanatory variable

ones – does the proportion of behaviour problems vary with ethnic group, with sex and with the interaction between ethnic group and sex? Now we find it difficult to answer these questions without a statistical model.

We can present the data in compressed form, corresponding to a 2×2×2 table, as follows.

| Sex | Ethnic group | With problem [count; (propn.)] | Total |
|---|---|---|---|
| 0 (boys) | 0 (white) | 19 (0.17) | 109 |
| 0 (boys) | 1 (black) | 33 (0.52) | 63 |
| 1 (girls) | 0 (white) | 20 (0.19) | 103 |
| 1 (girls) | 1 (black) | 18 (0.29) | 63 |

The first two rows are the same as Table 5.1, but we have added the data for girls, essentially as a further 2×2 table.

Our initial model, without the sex ethnic group interaction, is:

$$\text{logit } P_i = a + b_1 x_{1i} + b_2 x_{2i} \quad (5.16)$$

where $x_1$ is sex and $x_2$ is ethnic group.

The estimates are:

$\hat{a} = -1.3 \ (0.21)$

$\hat{b}_1 = -0.42 \ (0.26)$

$\hat{b}_2 = 1.12 \ (0.26)$

However, the residual deviance is 4.95 with one degree of freedom and this corresponds to the effect of the interaction between ethnic group and sex on logit $P_i$. This is statistically significant – $p<0.03$ – and so we cannot sensibly interpret the main effects on their own. When we look at the proportions in the data matrix above, we see that the ethnic group effect is much more marked for boys than it is for girls. Teachers' perceptions of behaviour problems are influenced not so much by ethnic group on its own, as we might have inferred from our analysis of Table 5.1, as by ethnic group and sex *in combination*.

What we have here is an example of what is sometimes called *Simpson's Paradox*, which states that an association present in a two-way table disappears, or changes or is even in the opposite direction, when the table is broken down by a third variable. It shows the importance of collecting all the data relevant to a particular problem and thus moving from the analysis of two-way tables to the analysis of multi-way tables.

Let us now turn to an example where some of the explanatory variables are ordered (dataset 5.1 on the disk). The data refer to the educational plans of a cohort of 4991 Wisconsin high school male pupils and come from Sewell and Shah (1968). They consist of about a one third random sample of all 1957 seniors in Wisconsin and have been re-analysed by, among others, Cox and Snell (1981). Our binary response is 'college plans' and there are three explanatory variables: parental encouragement (low, high), socio-economic status (a four-point scale from low (=1) to high (=4)) and IQ (also a four-point scale from low to high here, although measured originally on a continuous scale). The data are given in Table 5.4. They show that the proportion saying 'yes' tends to rise with IQ and socio-economic status as expected, and to be higher when parental encouragement is high.

## 86  Two by Two Tables and Beyond

**Table 5.4** College plans by IQ, parental encouragement and socio-economic status (counts and proportions saying 'Yes')

| IQ | College plans | Parental encouragement | Socio-economic status 1 | 2 | 3 | 4 |
|---|---|---|---|---|---|---|
| 1 | Yes | Low  | 4  [0.01] | 2   [0.01] | 8   [0.05] | 4   [0.08] |
|   |     | High | 13 [0.17] | 27  [0.24] | 47  [0.34] | 39  [0.41] |
|   | No  | Low  | 349       | 232        | 166        | 48         |
|   |     | High | 64        | 84         | 91         | 57         |
| 2 | Yes | Low  | 9  [0.04] | 7   [0.03] | 6   [0.05] | 5   [0.10] |
|   |     | High | 33 [0.31] | 64  [0.40] | 74  [0.40] | 123 [0.58] |
|   | No  | Low  | 207       | 201        | 120        | 47         |
|   |     | High | 72        | 95         | 110        | 90         |
| 3 | Yes | Low  | 12 [0.09] | 12  [0.09] | 17 [ 0.16] | 9   [0.18] |
|   |     | High | 38 [0.41] | 93  [0.50] | 148 [0.60] | 224 [0.76] |
|   | No  | Low  | 126       | 115        | 92         | 41         |
|   |     | High | 54        | 92         | 100        | 65         |
| 4 | Yes | Low  | 10 [0.13] | 17  [0.18] | 6   [0.12] | 8   [0.32] |
|   |     | High | 49 [0.53] | 119 [0.67] | 198 [0.73] | 414 [0.88] |
|   | No  | Low  | 67        | 79         | 42         | 17         |
|   |     | High | 43        | 59         | 73         | 54         |

We start with the following model:

$$\text{logit } P_i = a + b_1 x_{1i} + b_2 x_{2i} + b_3 x_{3i} \tag{5.17}$$

where $P_i$ is the probability of planning to go to college, $x_1$ is parental encouragement (0=low; 1=high), $x_2$ is socio-economic status which we will treat as ordered with integer scores from 1 to 4, and $x_3$ is IQ, also treated as ordered in the same way. (In the following example, we will see how to incorporate a continuously measured explanatory variable in the model.) Note that, as well as making assumptions about order, we do not include any interactions between the explanatory variables at this stage.

The model fits well; the residual deviance is 35.7 with 28 df ($p>0.15$). The estimates are:

$\hat{a} = -4.1$ (0.15) – the constant term

$\hat{b}_1 = 2.5$ (0.10) – the effect of parental encouragement; the odds of planning to go to college are multiplied by exp(2.5) or about 12 fold when parental encouragement is high.

$\hat{b}_2 = 0.47$ (0.037) – the linear effect of socio-economic status; the odds of planning to go to college are multiplied by 1.6 for each step up the status ladder.

$\hat{b}_3 = 0.67$ (0.037) – the effect of IQ; the odds of planning to go to college are just about doubled for each step up the observed IQ scale.

Just as in multiple regression, the sizes of each of these effects are for the variable in question when the values of the other two variables are held constant.

Up to now, we have had very little opportunity to carry out any model checking, basically because the residual degrees of freedom have been so small.

## 5.5 Logistic regression with more than one explanatory variable

However, there are possibilities for checking here. First, we can test whether the addition of all the non-linear components of socio-economic status and IQ improves the fit. That is, we treat socio-economic status and IQ as a set of dummies. (In Cox and Snell's analysis, these non-linear components were included.) We find a reduction of 10.4 in the deviance which, for four extra degrees of freedom, is statistically significant at the 0.04 level. However, given that the simpler model fitted well, and given its ease of interpretation, there are no strong grounds for including any non-linear terms. We can also obtain 'adjusted' residuals – adjusted for the fact that our response has a Binomial distribution – which should be close to Normality. The Q-Q plot, explained in Chapter 2, is given in Figure 5.2 and supports the Normality assumption. Also, the largest absolute value of the adjusted residuals is 2.82, not an unusual value in a sample of 64. In other words, our model stands up well to these checks.

We should note that the sample for this study was large. Even so, some of the cell counts are small. Constructing multi-way tables can quickly lead to an explosion of cells and not enough data to fill them. Consequently, we may be restricted as to the kinds of analyses we would like to do by the size of our sample.

We now give an example where one of the explanatory variables is genuinely continuous. The data – dataset 5.2 on the disk – come from a study of children's educational activities outside school, described in Plewis *et al.* (1990). The pupil sample of 230 Year one pupils was clustered into 19 schools, but we ignore the hierarchical structure here. The response is whether a child was a 'good' reader, this being defined on the basis of a reading test score. (In the original analysis, the reading test was treated as continuous; it is dichotomized here only for illustration. It is not usually good practice to throw away information by reducing a continuous measure to a binary one.) The explanatory variables are mother's education with three ordered levels (none, O-level, A-level and above) and the amount of reading aloud (in minutes) the child did at home the previous week, which is continuous although very much skewed towards zero. Our data file now has a row for each individual rather than a row for each cell of the multi-way

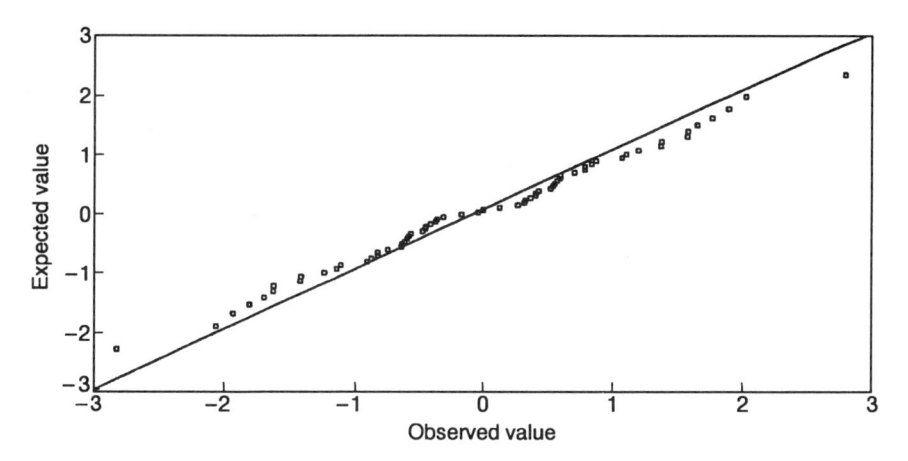

**Figure 5.2** Q-Q plot of standardized residuals

table created from cross-tabulating the response by all of the explanatory variables. We model the probability of being a good reader (the sample proportion is 13%), and we use the logistic regression option in SPSS.

As before, our model is:

$$\text{logit } P_i = a + b_1 x_{1i} + b_2 x_{2i}$$

where $P$ is the probability of being a good reader, $x_1$ is time spent reading aloud and $x_2$ is mother's education.

The inclusion of these two explanatory variables improves the fit (reduces the deviance) by 10.1, $p<0.01$. However, when we check our model, we find that there is one influential observation which has a value of Cook's $D$ (see p. 24) of 0.8, much greater than any other observation. The amount of time spent reading for this observation is 540 minutes per week, far greater than for any other observation. Excluding this observation, we find the reduction in deviance is 12.1 (rather than 10.1) and the parameter estimates and standard errors are:

$\hat{a} = -3.1$ (0.46);

$\hat{b}_1 = 0.0069$ (0.0032), so for every 10 minutes increase in reading aloud, the odds of being a good reader are multiplied by $\exp(0.0069 \times 10) = 1.07$;

$\hat{b}_2 = 0.83$ (0.33), so for each step up the education ladder, the odds of being a good reader grow by 2.3.

How good is our model now? First, we consider whether the assumption of equal effects as mothers move up their education scale is reasonable. We find it is because the deviance is hardly reduced at all by replacing mother's education with two dummy variables. Second, we can check how well the model predicts group membership, in our case good and less good readers. If the predicted probability for a pupil is less than 0.5, then this pupil is assigned to the category of less good readers, if greater than 0.5 then assignment is to the good reader category. We find that prediction of the less good readers, often called *specificity*, is perfect but that prediction of the good readers – *sensitivity* – is not. In fact, the model assigns all 23 pupils observed to be good readers to the less good category. Clearly, other explanatory variables are needed to understand better what makes a good reader at this age. For this reason, we cannot expect a Q-Q plot of the residuals to be close to a Normal distribution here because the predicted values for the good readers (coded 1) are close to zero. Hence, all the residuals for this group will be both positive and some distance from zero.

Finally, we must remember that our data have a hierarchical structure (pupils within schools) which we have ignored, not unreasonably as it happens because the between school variance for the response is small. However, a two-level logistic regression analysis might have produced estimates with slightly higher standard errors. A detailed treatment of multilevel models for categorical data is beyond the scope of this book, but some discussion of this topic can be found in the final chapter.

## 5.6 Methods for dependent samples

Up to now, we have been concerned with methods and models for data which could have been generated from a single random sample, or from a set of independent samples from different groups such as boys and girls. This means we do not yet have ways of dealing with data generated by repeated measurement, which create dependent samples. Repeated measurement generates *square tables* and square tables form an interesting special class of contingency tables. As well as arising when the same variable is measured on the same sample on two occasions, they can also arise from inter-rater reliability studies and from 'family' studies when we might be interested in, say, the reading problems of first and second born children.

Let us first consider the repeated measurement of a binary variable. The same ideas can be applied to 'family' studies. Table 5.5 gives an example: a sample of 149 children rated by their mothers, one year apart, on whether they were doing as well at school as they could.

A comparison of the marginal probabilities shows an increase of 10% in the proportion of mothers believing their child was doing well – from 0.46 to 0.56. However, this level of net or aggregate change might just be a chance occurrence and different samples might have given different patterns of change. The only information about change in the body of Table 5.5 comes from the off-diagonal cells – the individual changes from 'no' to 'yes' (31 children) and 'yes' to 'no' (17 children). These individual changes would have to be equal in the population for there to be no aggregate change, and the table would then be symmetric. We can test whether a two by two table is symmetric by using what has become known as McNemar's test:

$$M^2 = (b-c)^2/(b+c) \quad (5.18)$$

where $b$ and $c$ are the off-diagonal cells of the two by two table as before (see p. 78). $M^2$ has a $\chi^2$ distribution under the hypothesis of symmetry, with one degree of freedom. It takes the value 4.08 for Table 5.5 which is statistically significant at the 5% level. In other words, these data are consistent with a change in the proportion of children seen by their mothers as doing well between Year one and Year two; the increase of 10% from Year one to Year two is more than we would expect by chance.

It is important to realize that $M^2$ is a different statistic from $X^2$, and is used for testing a quite different hypothesis. With $M^2$ we are testing a hypothesis about symmetry and hence about aggregate change in two dependent samples, with $X^2$ we are testing whether there is an association between two variables. We could apply the methods of Section 5.2 to Table 5.5 and we would find that the two variables are strongly associated ($\hat{Q}=0.66$), but this would tell us nothing about change.

Let us now consider square tables generated by inter-rater reliability

**Table 5.5** Mothers' assessment of their children's school performance, Years one and two

|  |  | Doing well, Year two |  |  |
|---|---|---|---|---|
|  |  | No | Yes | Total |
| Doing well, Year one | No | 49 | 31 | 80 (0.54) |
|  | Yes | 17 | 52 | 69 (0.46) |
|  | Total | 66 (0.44) | 83 (0.56) | 149 (1) |

studies, when we want to know how well two raters agree. Association is not the same as agreement, and it is agreement which is relevant when we look at how well different raters or observers regard ostensibly the same phenomenon. Table 5.6 is an extreme example of a two by two table in which there is perfect negative association ($Q = -1$) but the two raters do not agree on any of the cases. In other words, a 'no' from rater one always predicts a 'yes' from rater two – and vice-versa – but the raters do not agree on how to rate any of the subjects.

Consider the data in Table 5.7. This shows how two raters of a sample of the same children – mothers and teachers – view their achievements at school. We can see that mothers are more negative overall than teachers, and that there is a fair degree of disagreement. How can we measure this agreement? A simple-minded way is just to add up the cells on the main diagonal and divide by the sample size to give percent agreement, and this statistic is often quoted in the literature. For Table 5.7, we get 58% agreement this way. Percent agreement is, however, a poor measure because it ignores the fact that raters can agree by chance, and chance agreement can be a substantial part of total agreement, especially when one category is much more prevalent than the other.

For example, suppose the prevalence of 'special needs' pupils is 10%. Also, suppose that two raters are blindfolded and so merely guess the classification of 100 pupils, but they maintain a 10% prevalence overall. Then, on average, they would each classify 81 of the pupils as 'not SN' and one as 'SN' and so they would agree on 82% of the pupils just by chance, as shown in Table 5.8.

Table 5.6  Perfect association and no agreement

|  |  | Rater one |  |  |
|---|---|---|---|---|
|  |  | No | Yes | Total |
| Rater two | No | 0 | 10 | 10 |
|  | Yes | 10 | 0 | 10 |
|  | Total | 10 | 10 | 20 |

Table 5.7  Child's achievement rated by mothers and teachers

| Child achievement as rated by: |  | Mother |  |  |
|---|---|---|---|---|
|  |  | Positive | Negative | Total |
| Teacher | Positive | 69 | 46 | 115 |
|  | Negative | 17 | 18 | 35 |
|  | Total | 86 | 64 | 150 |

Table 5.8  Chance agreement

|  |  | Rater one |  |  |
|---|---|---|---|---|
|  |  | Not SN | SN | Total |
| Rater two | Not SN | 81 | 9 | 90 |
|  | SN | 9 | 1 | 10 |
|  | Total | 90 | 10 | 100 |

A more appropriate measure of agreement is kappa, because it corrects for chance agreement. Kappa is defined as:

(Sum of observed cell sizes on the diagonal – sum of expected cell sizes on the diagonal given independence between observers)/(total sample – sum of expected cell sizes on the diagonal given independence )

which can be written more formally as:

$$\kappa = (\Sigma n_{ii} - \Sigma n_{i.} n_{.i}/n_{..})/(n_{..} - \Sigma n_{i.} n_{.i}/n_{..}) \quad (5.19)$$

where $n_{ii}$ are the cells on the diagonal, $n_{..}$ is the total sample and $n_{i.}$ and $n_{.i}$ are the row and column totals. In other words, both the numerator and the denominator are corrected for chance agreement.

Kappa and its standard error can be obtained as part of the SPSS CROSSTABS procedure. For Table 5.7, the estimate of kappa is 0.09 with a standard error of 0.08. This is clearly very small, and much smaller than the percent agreement. It shows how little agreement there is between mothers and teachers for this variable, and how important it is to correct for chance when measuring agreement.

## 5.7 Concluding remarks

We have seen how to describe and model data which come as proportions and probabilities from a single sample, or from independent samples. We have also seen how to summarize data in square tables generated from dependent samples, but we have not yet seen how to model such data. Rather than overload this chapter, modelling change and agreement are discussed in terms of responses with more than two categories, which is the subject of the next chapter.

## Exercises

The answers to exercises marked '#' can be found at the end of the book.

1  A random sample of 50 mothers each with a 12 year old child, living in London, are interviewed about their child's attitudes to homework. Half the mothers report that they find their child's attitude to homework to be a problem. The same sample of mothers is interviewed again two years later. Ten of the mothers report that their child's attitude to homework is a problem at the second interview. Eight mothers reported that their child's attitude to homework was a problem at both occasions.

Also, a random sample of 60 mothers of 14 year old children living in New York are asked the same question about attitudes to homework. A quarter of this sample reported problems with their child's attitude to homework.

Construct tables from these data in order to answer the following questions.
   (a) How much change in London mothers' perceptions of homework problems was there from 12 to 14 years, and was it more than you would expect by chance?
   (b) What do the data tell us about differences in the perceptions about

homework of London and New York mothers of 14 year olds?
(#)

2  A community is served by two schools, A and B. The following data are obtained about the number of pupils expelled from school over a school year.

| Outcome | Father employed | | Father not employed | |
|---|---|---|---|---|
| | School A | School B | School A | School B |
| Not expelled | 594 | 492 | 1443 | 192 |
| Expelled | 6 | 8 | 57 | 8 |

What is the association between outcome and school, regardless of father's employment status?

What is the association between outcome and school for each of the two employment statuses separately?

Comment carefully on your findings.

3  Using dataset 5.1 on the disk, take parental encouragement as the response and consider its relationship to socio-economic status and IQ.

4  Using dataset 5.2 on the disk, make a practicable categorization of the explanatory variable 'time spent reading aloud' and then fit the model described in Section 5.5. Compare the results with those given in Section 5.5.

5  Using dataset 5.2 on the disk, fit a model relating the probability of being a good reader to sex and time spent reading aloud, treated as a continuous variable. Is there any need in the model for the interaction between sex and time spent reading aloud?
(#)

# 6
# Larger Contingency Tables: Modelling Categorical Responses

## 6.1 Introduction

We were exclusively concerned in the previous chapter with binary responses. In this chapter, methods for the analysis of responses with more than two categories are presented. These responses might be unordered – type of educational handicap, for example – but are more likely to be ordered. Often these ordered responses are a rather crude representation of a continuous underlying, or latent, variable. Teacher assessments of pupils' ability on, say, a four-point scale is one example of this kind of variable. Despite the coarse-grained nature of measures of this kind, we should always try to extract as much information as we can from our data, by using the ordering in our analyses.

The chapter starts with a brief introduction to larger two-way tables, and shows how we can model associations between a set of categorical variables using log-linear models. We then move on to thinking about how we can represent order in categorical variables, and how we can model responses of this kind. Finally, we extend the ideas on change and agreement introduced in the previous chapter to variables with more than two categories, and show how we can model them.

## 6.2 Larger two-way tables

Table 6.1 is a contingency table relating teacher expectations for reading and mathematics at the beginning of Year one, obtained by interviewing teachers in 33 schools about a sample of 450 pupils. The data were collected in the Tizard *et al.* (1988) study (see p. 57). There are three rows and three columns, representing the ordered three-point scales used. We will continue to label rows and columns by subscripts $i$ and $j$, but will use $I$ and $J$ to represent the number of rows and columns. Here, $I=J=3$. In this case, there is no clear response variable and so, initially, we want to measure the association between the two sets of expectations.

94  *Larger Contingency Tables*

**Table 6.1** Teacher expectations, reading and mathematics (counts and marginal percentages)

| Mathematics | Reading | | | |
| --- | --- | --- | --- | --- |
| | Below | Average | Above | Total |
| Below | 102 | 13 | 0 | 115 (26%) |
| Average | 35 | 141 | 20 | 196 (44%) |
| Above | 3 | 18 | 118 | 139 (31%) |
| Total | 140 (31%) | 172 (38%) | 138 (31%) | 450 (100%) |

The fact that 80% of the sample is to be found down the main diagonal of the table, and that there are very few pupils in the two corner cells, suggests that there is a marked association between the two sets of expectations. Kendall's tau-$c$ is again a suitable measure of association (see Section 5.3); the value of $\tau_c$ for Table 6.1 is 0.77 with a standard error of 0.02, indicating a strong degree of association.

Clearly, the association here is far greater than we would expect by chance. If we were uncertain about this, we could use the Mantel–Haenszel test introduced in Section 5.3, and extended to situations where both variables are ordered. For Table 6.1, the Mantel–Haenszel chi-square is 301, with one degree of freedom, giving a very low $p$-value. We should remember that Kendall's $\tau_c$ makes no assumptions about the two scales other than that they are ordered. The Mantel–Haenszel test, on the other hand, is based on the assumption that scores can be assigned both to the row and to the column variables, and here the scores were the equally spaced integers from one up to $I$ (i.e. 3), and one up to $J$ (i.e. 3).

In Chapter 5 we saw how to test this assumption of equal spacing for the effect of an ordered categorical explanatory variable on a binary response. We would like to be able to do the same in situations like those represented by Table 6.1, where it is not possible to say that one of the variables is the response and the other explanatory. To do this, it is helpful to introduce a new class of statistical model, the log-linear model. The log-linear model is a flexible model which enables us to model the associations between two or more variables, which could be ordered or unordered, categorical or binary.

## 6.3 Log-linear models for modelling association

In Chapter 5, we modelled binary responses using logistic regression – a logit, or log odds link between the response and the explanatory variables, and Binomial errors. Later in this chapter we will model ordered categorical responses using extensions of log odds for the response. However, when we have two or more variables with no obvious response, then we model the cell counts using a log link and Poisson errors. Like the Binomial distribution, the mean and variance of the Poisson distribution are not independent; in fact, they are equal. The log-linear model is therefore another kind of generalized linear model. We aim to use this model to simplify the association structure in a multi-way table by picking out the most important relations within it, and discarding the rest. A useful introduction to log-linear models for social data, based on the SPSS package, is given by Rose and Sullivan (1996), while Gilbert (1993) goes into more detail without being too technical.

Let us start with Table 6.1, initially ignoring the ordering in the two teacher expectation measures. The simplest log-linear model is one which assumes independence. Remember from Chapter 5 that the hypothesis of independence can be expressed as:

$$N_{ij}=n_i n_j/n_{..} \qquad (6.1)$$

where we now use $N_{ij}$ (rather than $e_{ij}$) as the expected count for cell $i, j$ and with $\Sigma_i n_{i.}$ (the row counts)$=\Sigma_j n_{.j}$ (the column counts)$=n_{..}$ (the sample size).

This is a multiplicative model, which we can transform into a computationally more convenient additive one by taking logs (again to base e). This is why we refer to the models as log-linear models. Thus, the independence model is:

$$\log_e N_{ij}=\log_e n_{i.}+\log_e n_{.j}-\log_e n_{..} \qquad (6.2)$$

which we can write more conveniently as:

$$\log N_{ij}=a+r_i^M+c_j^R \qquad (6.3)$$

where $a$ is a constant, $r_i^M$ is the main effect of the row variable, teacher expectations for mathematics ($i=1,2,3$); and $c_j^R$ is the main effect for the column variable, teacher expectations for reading ($j=1,2,3$). These two main effects are the sets of parameters which fix the expected marginal counts to equal the observed marginal counts of the table and they must always be in the model. Because the row and column marginal counts, $n_{i.}$ and $n_{.j}$, must both sum to the same total ($n_{..}$, the sample size), the number of degrees of freedom for each main effect is $I-1$ and $J-1$, which are two and two here. The most convenient way of ensuring that this constraint on the margins is satisfied is to set one of the $r_i^M$ and one of the $c_j^R$ to zero. We can think of the $r_i^M$ and $c_j^R$ as parameters attached to two sets of dummy variables (see Chapter 2).

There are $I$ times $J$ cells in a two-way table (=nine in Table 6.1). The above model estimates 1 (the constant term, $a$)$+(I-1)+(J-1)$ parameters (the two main effects), leaving $IJ-[1+(I-1)+(J-1)]$ degrees of freedom or $(I-1)(J-1)=4$ in this case. These are, of course, the number of degrees of freedom for the usual chi-square test of independence described in Section 5.2. When we fit the above model to a two-way table, we are essentially testing for independence. The LOGLINEAR menu in SPSS allows different kinds of log-linear models to be fitted using the maximum likelihood method described in, for example, McCullagh and Nelder (1989). However, the details of the estimation method need not concern us here.

It is important to remember that we use a log transformation of the *expected* cell sizes (i.e. $N_{ij}$) in log-linear models, just as we transformed the probabilities rather than the observed proportions in the models for binary responses in Chapter 5. Our approach here is quite different from using a log transformation on a continuous response, as we did in Chapter 2 to try to meet the assumptions of the linear regression model. One of the reasons for using a log link is that many of the statistics calculated from contingency tables are expressed in terms of multiplication and division of cell counts and marginal totals, for example the odds ratio (p. 78) and, as we have just seen, the hypothesis of independence.

As we already know, this simplest log-linear model with just a constant and

## 96  Larger Contingency Tables

the two main effects fits badly to the data in Table 6.1. The residual deviance is 463 with four degrees of freedom. Clearly, the two sets of teacher expectations are associated, and there is no support for the hypothesis of independence. In log-linear models, unlike multiple and logistic regression, associations are represented by interactions. Here, we represent the association between the two teacher expectations by a set of four interaction parameters, $A_{ij}^{RM}$. If we fit all these interaction parameters, we end up with a saturated model with no residual degrees of freedom, and hence with the expected and observed counts exactly the same.

Let us instead use the ordering in the two measures to see whether we can get a good fit without fitting all four interaction parameters. In other words, can we find a parsimonious way of representing the interaction? One approach we can adopt is to score each expectations measure as 1,2,3, calling $x_i$ the score for mathematics and $y_j$ the score for reading. Then we fit a term known as a *linear by linear* association. We do this by creating a new variable, $z_{ij}$, which is just the product of the two scored variables, so $z_{ij}=x_i y_j$ and takes the integer values 1, 2 (twice), 3 (twice), 4, 6 (twice) and 9. Our model is now:

$$\log N_{ij} = a + r_i^M + c_j^R + b\, z_{ij} \tag{6.4}$$

where $b$ is the coefficient which gives the size of the linear by linear association.

This model fits much better than the previous one; the deviance falls by 451 for the loss of just one degree of freedom from estimating $b$. The estimate of $b$ is 3.36 with a s.e. of 0.24, and shows that the two sets of expectations are positively associated.

Despite the improvement in fit, the residual deviance is still 12.4 with three df, more than we would expect by chance. This suggests that our search for a good model for these data should not end here. We can get some idea of why the model still does not fit well by looking at the adjusted residuals (introduced on p. 87). We find that these residuals are large for the 'average' and 'above average' diagonal cells. In other words, more pupils get the same expectations rating for the two subjects than the linear by linear association model predicts. A better model might be one which fixes the diagonal cells so that the observed and expected cell counts are equal and then looks at association solely for the off-diagonal cells. This is known as a *quasi-independence* model – see Exercise 1 at the end of this chapter.

Just as the advantages of regression and logistic regression were revealed when there was more than one explanatory variable, so the strengths of the log-linear approach are to be found when there are more than two variables altogether. In other words, when we have multi-way tables, rather than just an *I* by *J* table. So, suppose we break down Table 6.1 by sex, creating a three-way, *I* by *J* by *K* table. This is Table 6.2, which is 3×3×2, and so has 18 cells.

This leads us to a set of new questions:

(i)  Is there an association between sex and teacher expectations for mathematics, both before *and* after controlling for other associations in the table?
(ii) Is there an association between sex and teacher expectations for reading, again both before and after controlling for other associations in the table?
(iii) Does the association between the two teacher expectations measures (already established from the analysis of Table 6.1) vary by sex?

## 6.3 Log-linear models for modelling association

**Table 6.2** Teacher expectations, reading and mathematics by sex (counts and marginal percentages)

| Mathematics | | Reading | | | |
|---|---|---|---|---|---|
| | | Below | Average | Above | Total |
| Boys | Below | 62 | 5 | 0 | 67 (27%) |
| | Average | 30 | 66 | 7 | 103 (42%) |
| | Above | 3 | 11 | 62 | 76 (31%) |
| | Total | 95 (39%) | 82 (33%) | 69 (28%) | 246 (100%) |
| Girls | Below | 40 | 8 | 0 | 48 (24%) |
| | Average | 5 | 75 | 13 | 93 (46%) |
| | Above | 0 | 7 | 56 | 63 (31%) |
| | Total | 45 (22%) | 90 (44%) | 69 (34%) | 204 (100%) |

We should note that there are three cells with zero counts. These are sampling zeros, which would not necessarily occur with a different or a larger sample, and they do not prevent us from fitting log-linear models to the table. However, we should be aware that a substantial proportion of cells with counts less than five can lead to misleading values of the deviance, and hence to unsound conclusions about the goodness of fit of the model. Also, we cannot fit log-linear models if any of the marginal counts are zero.

There are a number of log-linear models which we might fit to the data in Table 6.2. We must include a main effect for sex to fix the totals for boys and girls, and so our basic model with no associations is now:

$$\log N_{ijk} = a + r_i^M + c_j^R + l_k^S \tag{6.5}$$

which has three subscripts corresponding to the rows ($i$:mathematics), columns ($j$:reading) and layers ($k$:sex), with $k$ taking the value 1 for boys and 2 for girls, and with $l_2 = 0$.

We find that this independence model does not fit well; the deviance is 492 with 12 df ($p < 0.001$). Next we see whether there is an association between mathematics expections and sex and so the model expands to:

$$\log N_{ijk} = a + r_i^M + c_j^R + l_k^S + A_{ik}^{MS} \tag{6.6}$$

This extra term results in a decrease in the deviance of just 1 for two (i.e. $(I-1)(K-1)$) extra degrees of freedom. So there does not appear to be an association between mathematics expectations and sex. Essentially, this is telling us that, if we collapse the table over the reading expectations variable, there is no association between mathematics expectations and sex.

Suppose, instead of $A_{ik}^{MS}$, we include the association between reading expectations and sex, giving us:

$$\log N_{ijk} = a + r_i^M + c_j^R + l_k^S + A_{jk}^{RS} \tag{6.7}$$

This model is an improvement over the basic model; the deviance is 477 with 10 df, an improvement of 15 for 2 df. When we collapse over mathematics expectations, there is an association between reading expectations and sex.

Does this association hold up after we have allowed for the association between the two expectations measures? First we fit all four interaction parameters for the association between the two sets of expectations:

$$\log N_{ijk} = a + r_i^M + c_j^R + l_k^S + A_{ij}^{MR} \tag{6.8}$$

The deviance for this model is 29.1 with eight df, a huge fall from the basic independence model, but still not a model that fits well. Now we add the interaction between sex and reading, $A_{jk}^{RS}$, to the model. We find this reduces the deviance by 14.7 for two df ($p<0.001$). However, the model still does not fit well – the residual deviance is 14.4 with six df ($p<0.03$) – so we bring back the interaction between sex and mathematics, $A_{ik}^{MS}$. Perhaps surprisingly, the fit of the model improves further; the deviance falls by 12.0 for two extra parameters (or df), much more than we would expect by chance. Moreover, the residual deviance is now very small – 2.37 with four df.

This small residual deviance implies that the answer to question (iii) on p. 96 is 'no'. This is because the variation in the association of the two expectations measures by sex is represented by the three-way interaction, which is now the only term missing from the model. Including the three-way interaction at this stage would improve the fit by 2.37 for the loss of the four (i.e. $(I-1)(J-1)(K-1)$) remaining df – in other words, hardly at all.

The evidence from the sequence of models we have fitted does support answers of 'yes' to both parts of question (ii). The answer to question (i) is an interesting one. There is no association between mathematics expectations and sex *before* we control for the other associations but there is *after* controlling for those associations. This is another example of Simpson's Paradox (see p. 85).

We might be able to simplify the model, and hence the answer to question (i), by using the ordering in the two expectations measures in their interactions with sex. We do this by replacing the two interactions, $A_{ik}^{MS}$ and $A_{jk}^{RS}$ both with two df, with two linear association terms, each having one df, which we allow to vary with sex. However, we retain all four interaction parameters for $A_{ij}^{MR}$, so we have:

$$\log N_{ijk} = a + r_i^M + c_j^R + l_k^S + A_{ij}^{MR} + b_{1k}x_i + b_{2k}y_j \tag{6.9}$$

where $x_i$ is the scored version of mathematics expectations and $y_j$ is the scored version of reading expectations (see p. 96), and $b_{1k}$ and $b_{2k}$ are coefficients which vary by sex ($k=1$ for boys, $k=2$ for girls). This model also fits well, with a residual deviance of 7.76 with six df. The increase in deviance from the model with the full interactions is 5.39 with two df which has a $p$-value of 0.07. So the balance of the evidence favours the simpler model which assumes a particular order. The two estimated coefficients are:

$\hat{b}_{11} = 0.83$ (0.25) ($b_{12}=0$)
$\hat{b}_{21} = -1.0$ (0.24) ($b_{22}=0$)

In other words, the positive coefficient for $b_{11}$ shows that, after controlling for the association between the two sets of expectations and for the association between reading and sex, the odds of getting a higher mathematics expectation are greater for boys than for girls. This finding is masked when we look just at the marginal distributions of mathematics expectations and sex in Table 6.2. On the other hand, the negative estimate for $b_{21}$ shows that the odds of getting a lower expectation for reading are greater for boys, after controlling for the other two associations.

Table 6.3 gives a summary of all seven models fitted to Table 6.2. We can only

## 6.3 Log-linear models for modelling association

**Table 6.3** Log-linear models fitted to Table 6.2

| Model | Deviance | df |
|---|---|---|
| 1. M + R + S | 492 | 12 |
| 2. M + R + S + M.S | 491 | 10 |
| 3. M + R + S + R.S | 477 | 10 |
| 4. M + R + S + M.R | 29.1 | 8 |
| 5. M + R + S + M.R + R.S | 14.4 | 6 |
| 6. M + R + S + M.R + R.S + M.S | 2.37 | 4 |
| 7. M + R + S + M.R + $R^L$.S + $M^L$.S | 7.76 | 6 |

*Notes:* Models 2, 3 and 4 are nested within model 1. Model 5 is nested within models 3 and 4. Model 6 is nested within models 5 and 7. Model 7 is nested within model 4.

make statements about improvements in goodness of fit for those models which are nested or contained within a simpler model, as indicated. If model A is *nested* within model B, and model B is nested within model C, then model A must also be nested within model C. However, we cannot directly compare the improvement in the fit of, say, model 4 over model 3 as model 3 includes a term (R.S) not in model 4.

Table 6.4 gives the expected or fitted counts from the model, together with the adjusted residuals. Comparing Tables 6.2 and 6.4, we see that the largest discrepancies are when both expectations are average. However, an adjusted residual of 2.03 for boys (and therefore –2.03 for girls because the observed and expected marginal totals must agree) is not unduly high. Thus, we have found a model that fits well to these data, and which raises some interesting questions about how teachers form expectations about their pupils. We might, in fact, have arrived at this point by a different route – see Exercise 2.

It is important to note that the inclusion of interactions in log-linear models should be guided by the same principles that were set out for regression models in Chapter 2. In other words, no two-way interactions can be included in a model unless the two corresponding main effects are included, no three-way interactions without the three corresponding two-way interactions, and so on.

This has been a necessarily brief introduction to log-linear modelling. The approach can be useful, as we have seen, particularly when we want to try to simplify a set of associations. We shall also use it again in Section 6.7, when we model agreement. However, models which make a clear distinction between the response and the explanatory variables tend to be more useful when we want to understand educational processes. As we saw in Chapter 5, logistic regression is one member of that class, and we now move on to consider others.

**Table 6.4** Fitted counts and adjusted residuals, model 7

| Mathematics | | Reading | | |
|---|---|---|---|---|
| | | Below | Average | Above |
| Boys | Below | 59.9 (0.77) | 4.36 (0.42) | 0.00 (0.00) |
| | Average | 26.8 (1.65) | 75.8 (–2.03) | 5.85 (0.67) |
| | Above | 2.65 (0.65) | 13.1 (–1.28) | 57.5 (1.64) |
| Girls | Below | 42.1 (–0.77) | 8.64 (–0.42) | 0.00 (0.00) |
| | Average | 8.19 (–1.65) | 65.2 (2.03) | 14.2 (–0.67) |
| | Above | 0.35 (–0.65) | 4.9 (1.28) | 60.5 (–1.64) |

## 6.4 Representing order in categorical responses

Assigning the integer values 1,2 and 3 to the three scale points, as we did with Table 6.1, does impose the rather severe assumption of equality of distance between each category. It is also an assumption which is difficult to use for a response in a statistical model because the distribution is bound not to be Normal. Some alternatives to this scoring method are shown in Table 6.5 (rows 1 and 2) for the marginal distribution of the reading expectations scale from Table 6.1. One possibility would be to use the cumulative probabilities (row 4) or their complement, the survival probabilities (row 5). However, our models for binary responses were based on odds rather than on probabilities. There are good theoretical and practical reasons for maintaining this link with odds for categorical responses, and so either *cumulative logits* (row 6) or logged *continuation odds* (row 7) might be better. For an ordered $K$ category response, there are $K-1$ cumulative logits and $K-1$ continuation odds.

Here, the two cumulative logits are defined as: (i) the log of the odds of being below average to being *at least* average (140/310), and (ii) the log of the odds of being at least average to being above average (312/138). Thus, we express cumulative logits as the log of the ratio of each cumulative probability to its complement, which is one minus the corresponding cumulative probability.

The two logged continuation odds are (i) the log of the odds of being at least average to being below average (310/140) (just the reciprocal of the first cumulative logit), and (ii) the log of the odds of being above average to being average (138/172). Continuation odds are sometimes called *continuation ratios*. They can be expressed as the ratio of the survival probability to the corresponding category probability. As we shall see, the choice between these two ways of representing ordered responses will depend partly on the meaning of the response variable, and partly on how well the model fits the data.

If we were to ignore the ordering in the reading expectations variable, or if we have an unordered categorical response, then we could use what are often called *multivariate logits*. In the final row of Table 6.5, these are given as (i) the log of the odds of being below average to being above average (140/138) and (ii) the log of the odds of being average to being above average (172/138). In other words, above average is the baseline category. However, because the ordering is irrelevant to the construction of multivariate logits, either of the other two categories could have been used as the baseline.

**Table 6.5** Representations of ordered categorical variables

|  | Below average | Average | Above average |
|---|---|---|---|
| Frequency | 140 | 172 | 138 |
| Probabilities | 0.311 | 0.382 | 0.307 |
| Cumulative frequencies | 140 | 312 | 450 |
| Cumulative probabilities | 0.311 | 0.693 | 1 |
| Survival probabilities | 0.689 | 0.307 | 0.00 |
| Cumulative logits |  | −0.79 | 0.82 |
| Logged continuation odds |  | 0.79 | −0.22 |
| Multivariate logits |  | 0.01 | 0.22 |

## 6.5 Modelling ordered responses

Let us now look at some data similar to those given in Table 6.2. However, this time we have data on 354 pupils from the same study, and we are interested in the relation between teacher expectations for reading (the ordered response) at the beginning of Year two and sex (a binary explanatory variable). Do teachers have higher expectations for boys' or for girls' reading at this stage of their schooling? Table 6.6 gives the data, together with the cumulative logits. We see that there are some differences between boys and girls, with the cumulative logits lower for girls, which means that expectations are higher for girls.

In this example (Table 6.6), we use cumulative logits for the response because it is possible, at least in principle, for pupils to move both up and down the expectations scale over time. Our second example will use logged continuation odds. We can write the model as:

$$\log[\text{Cu}P_{ik}/(1-\text{Cu}P_{ik})] = a_k + v_i^S \qquad (6.10)$$

where $\text{Cu}P_{ik}$ is the $k$th expected cumulative probability ($k=1...K-1$) for category $i$ of the explanatory variable, sex. In our case, $K$ is 3 and $I$ is 2 and so, with the usual constraints, $v_2$ is zero. The $a_k$ are constant terms which, in this case, give the expected cumulative logits averaged over the two sex categories. The important point to note about this model is that the effect of sex is assumed to be the same for each of the two cumulative logits. For this reason, the model is often known as a *proportional odds* model. We would not be exploiting the order in the response if this proportionality did not hold because we would then be implying that different orderings, or scalings, of the response are needed for the two sexes. Nevertheless, the proportionality assumption can be tested.

We cannot use SPSS to model ordered categorical responses. However, we can use the CATMOD procedure in SAS (see Appendix 6.1 for the relevant code). We should note that, for these kinds of models, CATMOD uses weighted least squares (WLS) rather than maximum likelihood as its estimating method. This need not concern us here but it does mean that, rather than describing fit in terms of the deviance, we use the phrase 'residual chi-square'.

Our estimate of $v_1$ is 0.57 (s.e.=0.20), which means that these cumulative logits for boys are 0.57 higher than they are for girls. Another way of putting this is to say that each of the two odds (below average to at least average, and at least average to above average) are exp(0.57)=1.77 higher for boys than for girls. The model fits well – the residual chi-square is 0.83 with one df – and so it is reasonable to conclude that there is a proportional sex effect for these two cumulative logits, with boys lower on the expectations scale than girls.

Table 6.6  Teacher expectations by sex

|         | Boys |     |                  | Girls |     |                  |
|---------|------|-----|------------------|-------|-----|------------------|
|         | n    | %   | Cumul. Logits    | n     | %   | Cumul. Logits    |
| Below   | 56   | 30  | −0.86            | 28    | 17  | −1.59            |
| Average | 69   | 37  |                  | 62    | 38  |                  |
| Above   | 64   | 34  | 0.67             | 75    | 46  | 0.18             |
| Total   | 189  | 100 |                  | 165   | 100 |                  |

## 102  Larger Contingency Tables

Our second example is based on a subset of data collected from three schools during a pilot study leading up to the introduction of national assessment in England and Wales. The full dataset is used in Section 6.6, and is described in more detail there. Here, we have data for spelling measured as levels on a three-point scale broken down by sex (Table 6.7). We see that girls get higher levels than boys and therefore their logged continuation odds are higher, especially for the first pair of continuation odds.

Because pupils can only move up the scale defined by these levels, which can be thought of as a series of hurdles, continuation odds are a more appropriate way of representing order.

We write the model for continuation odds as:

$$\log [\mathrm{Su}P_{ik}/P_{ik}] = a_k + v_i^S \qquad (6.11)$$

where $\mathrm{Su}P_{ik}$ is the $k$th survival probability (see Table 6.4), and $P_{ik}$ is the probability ($k=1...K-1$), for category $i$ of the explanatory variable. As before, $K$ is 3 and $I$ is 2, and again we assume that the effect of sex is the same for each logged continuation odds.

We can use CATMOD in SAS to model continuation odds, although we have to derive the responses ourselves within the procedure (see Appendix 6.1). The estimate of $v_1$ is $-0.96$ (s.e.$=0.40$) with $v_2$ set to zero, showing the continuation odds for boys are lower than for girls, or that boys are lower on the scale. For these data, the two continuation odds (at least average to below average, and above average to average) are $\exp(-0.96)=0.38$ times higher for boys than for girls, or 2.6 (1/0.38) times higher for girls than for boys. The residual chi-square is 3.16 with one df ($p>0.07$), suggesting a reasonable fit to the data.

However, we know the data come from three schools (with 33, 42 and 36 pupils) and so we put school into the model as two dummy variables. Our model is now:

$$\log [\mathrm{Su}P_{ijk}/P_{ijk}] = a_k + v_i^S + w_j^{Sc} \qquad (6.12)$$

and $w_j$ are the effects for school ($w_3=0$). We assume for now that the sex effect does not vary by school. We find that the school effect is small for these data (which are given in Appendix 6.1 within the SAS code) with a chi-square statistic of 3.56 with two df ($p>0.16$). The sex effect increases a little from the previous model, to $-1.1$ (s.e.$=0.41$) and the residual chi-square is 10.4 with seven df ($p>0.16$) and so the model fits well. There is no evidence to suppose that the continuation odds vary by school nor that the sex effect varies by school. We need to bear in mind, however, that the cell sizes within schools are small and therefore

**Table 6.7** Spelling levels by sex

| Level | Boys |     |                    | Girls |     |                    |
|-------|------|-----|--------------------|-------|-----|--------------------|
|       | n    | %   | Log Cont. Odds     | n     | %   | Log Cont. Odds     |
| 1     | 20   | 34  | 0.64               | 5     | 9   | 2.26               |
| 2     | 32   | 55  |                    | 39    | 74  |                    |
| 3     | 6    | 10  | −1.67              | 9     | 17  | −1.47              |
| Total | 58   | 100 |                    | 53    | 100 |                    |

our tests of the school effect, and the interaction of school with sex, have rather low power.

We have reached a rather similar position with these data as we did in Chapter 2 when we analysed a subset of data from three schools. In Chapter 3, we went on to do a full analysis of those data using all schools in a multilevel analysis. In principle, we could do the same here. However, multilevel analysis of categorical responses is complicated and we do not pursue it further at this point. Some of the issues are discussed in Chapter 8.

## 6.6 Modelling change

In Chapter 5 (Section 5.6), we saw how to look at binary responses measured on two occasions on the same sample. We extend those ideas to categorical responses here, and hence to larger square tables, and we introduce ways of modelling change in categorical responses.

Consider Table 6.8, which shows teacher expectations for mathematics collected from the same sample of 353 pupils at the beginning of Years one and two. We see that about 60% of the sample get the same expectation and that very few pupils get ratings which are more than one level apart. When we look at the marginal distributions, there appears to be only a little overall change, with a slight shift towards the 'above average' category.

We can extend the McNemar test for symmetry (p. 89) to square tables with more than two categories. If we find the table is symmetrical then we can conclude that there is no aggregate or marginal change from Year one to Year two. However, the converse does not hold once we move away from binary responses; lack of symmetry does not necessarily imply marginal change, as we see in Table 6.9. Often, the case of no marginal change is referred to as *marginal homogeneity*.

In the general case, the McNemar statistic is:

$$M^2 = \Sigma(n_{ij} - n_{ji})^2/(n_{ij} + n_{ji}) \quad \text{for } i \neq j \qquad (6.13)$$

**Table 6.8** Teacher expectations, mathematics, Years one and two

| Year one | Year two | | | |
| --- | --- | --- | --- | --- |
| | Below average | Average | Above average | Total |
| Below average | 53 | 27 | 5 | 85 (24%) |
| Average | 24 | 84 | 46 | 154 (44%) |
| Above average | 2 | 38 | 74 | 114 (32%) |
| Total | 79 (22%) | 149 (42%) | 125 (35%) | 353 (100%) |

**Table 6.9** A square table with marginal homogeneity but not symmetry

| | 1 | 2 | 3 | Total |
| --- | --- | --- | --- | --- |
| 1 | 40 | 2 | 8 | 50 |
| 2 | 8 | 30 | 12 | 50 |
| 3 | 2 | 18 | 0 | 20 |
| Total | 50 | 50 | 20 | 120 |

and this has a chi-square distribution with $r(r-1)/2$ degrees of freedom where $r$ is the number of rows (and columns). For Table 6.8, $M^2$ is 2.23 with three df and so Table 6.8 is clearly consistent with symmetry, and hence marginal homogeneity, in the population. For Table 6.9, $M^2$ is 8.4 with three df, $p<0.04$, and so it is not symmetric.

Table 6.10 is the same as Table 6.9 but for reading. Here we see more change in the form of upward movement and the hypothesis of symmetry is rejected as $M^2 = 11.5$ with three df ($p<0.01$). However, this does not tell us whether the observed marginal change is more than we would expect by chance.

There are no easily applicable tests of marginal homogeneity for tables like Table 6.10. However, we can model the marginal probabilities and this method is described below. The SAS code, using CATMOD, is set out in Appendix 6.1. We find that the test of the hypothesis that the effect of year is zero, which is the test of marginal homogeneity, has a chi-square statistic of 11.5 with two df. Therefore, the observed amount of aggregate change is more than just a chance difference ($p<0.004$). The chi-square statistic for marginal homogeneity is very close to the one obtained above for symmetry, suggesting that all the departure from symmetry arises from differences in the margins. Expectations for reading are higher in Year two than in Year one.

We might also ask whether there are different amounts of aggregate change for boys and girls. Table 6.11 gives the data broken down by sex. Is there more change for boys (the top figure in each cell) than there is for girls (the lower figure)? We could still have asked this question if there had been no overall aggregate change, because boys could have changed in one direction and girls in the opposite direction.

We are now in a rather similar situation to the one we met in Chapter 2, when we related change in mathematics attainment to sex and found that there were

**Table 6.10**  Teacher expectations, reading, Years one and two

| Year one | Year two | | | |
|---|---|---|---|---|
| | Below average | Average | Above average | Total |
| Below average | 58 | 37 | 6 | 101 (29%) |
| Average | 25 | 63 | 50 | 138 (39%) |
| Above average | 1 | 29 | 84 | 114 (32%) |
| Total | 84 (24%) | 129 (37%) | 140 (40%) | 353 (100%) |

**Table 6.11**  Teacher expectations, reading, Years one and two by sex

| Year one | Year two | | | |
|---|---|---|---|---|
| | Below average | Average | Above average | Total |
| Below average | 39 | 28 | 0 | 67 (35%) |
| | 19 | 9 | 6 | 34 (21%) |
| Average | 16 | 25 | 24 | 65 (34%) |
| | 9 | 38 | 26 | 73 (45%) |
| Above average | 1 | 16 | 40 | 57 (30%) |
| | 0 | 13 | 44 | 57 (35%) |
| Total | 56 (30%) | 69 (37%) | 64 (34%) | 189 (100%) |
| | 28 (17%) | 60 (37%) | 76 (46%) | 164 (100%) |

two approaches to the analysis, a conditional, or regression, model and an unconditional model based on difference scores. We are in the same position here.

The conditional model asks whether there is any association between expectations for Year two and sex, *after* controlling for the association between Year one and Year two expectations. This is the conditional approach to the question of whether the change in expectations is related to sex. The model is, in essence, the same as those used in Section 6.5: we relate the cumulative logits for Year two expectations to Year one expectations and sex. In other words:

$$\log[\text{Cu}P_{ijk}/(1-\text{Cu}P_{ijk})] = a_k + v_i^S + w_j^{Y1} \tag{6.14}$$

where $i$ stands for sex so $v_2=0$, $j$ stands for Year one expectations so $w_3=0$ and $k$ is 1 or 2 because there are three categories and thus two cumulative logits. Our interest focuses on the estimate of $v_1$.

The model fits well enough; the residual chi-square is 12.6 with 7 df ($p>0.08$). Therefore, there is no evidence to suppose that the effect of sex varies by the level of Year one expectations nor that the proportional odds assumption is unreasonable. The estimate of $v_1$ is $-0.27$ with a standard error of 0.22 which is consistent with a value of zero in the population. In other words, having allowed for the effect of Year one expectations on Year two expectations (which is of course substantial), there is no association between sex and Year two expectations. This in turn means that there is no *relative* change in expectations from a conditional model.

For the unconditional model, the response is the cumulative logits for the marginal probabilities:

$$\log[\text{CuMa}P_{ijk}/(1-\text{CuMa}P_{ijk})] = b_k + e_i^S + f_j^Y \tag{6.15}$$

where $\text{CuMa}P_{ijk}$ is the cumulative marginal probability for sex $i$ ($i=1,2$) and year $j$ ($j=1,2$). The observed marginal percentages, from which the cumulative marginal percentages can easily be derived, are given in Table 6.11. If this main effects model fits well then we can conclude that the interaction between sex and year is not important, which in turn means that the unconditional change is the same for boys and girls.

Fitting this model is a little complicated using CATMOD and the details are given in Appendix 6.1. However, the main effects model does fit well ($\chi^2=1.93$, 2 df, $p>0.3$) and so, in this case (but not always, as shown by Plewis, 1981), the conclusions from the conditional and unconditional models are the same. The test of whether the main effect for year (i.e. $f_1$) is zero is just the test for marginal homogeneity discussed earlier.

## 6.7 Modelling agreement

In the previous chapter (Section 5.6), we became aware of the difference between association and agreement, and saw how to measure agreement for two by two tables, using the kappa coefficient. When categories are not ordered, the extension of kappa to larger tables is just a straightforward extension of the formula given on p. 91. But when there is ordering, then we should use that information because disagreements of just one category are clearly less severe than disagreements of more than one category. This leads first to the notion of a weighted version of kappa ($\kappa_w$), as described by Fleiss (1981):

$$\kappa_w = (n_{..}\Sigma w_{ij} n_{ij} - \Sigma w_{ij} n_{i.} n_{.j})/(n_{..}^2 - \Sigma w_{ij} n_{i.} n_{.j}) \quad (6.16)$$

where the $w_{ij}$ are weights with the following properties: $w_{ii} = 1$, $0 \leq w_{ij} \leq 1$ ($i \neq j$), $w_{ij} = w_{ji}$ and $w_{ij} > w_{i,j+k}$ ($k > 0$). In other words, disagreements of one category in either direction are given equal weight and the greater the distance between the two ratings, the smaller the weight. The unweighted version of kappa has $w_{ij} = 0$ for all $i \neq j$. The choice of weights within these constraints is arbitrary. One possibility for $w_{ij}$ is:

$$\text{(a)} \quad w_{ij} = 1 - (i - j)^2/(I - 1)^2 \quad (6.17)$$

where I is the number of rows and columns. Another is:

$$\text{(b)} \quad w_{ij} = 1 - |i - j|/(I - 1) \quad (6.18)$$

where $|i - j|$ is the absolute difference between $i$ and $j$, the number of categories between the ratings of the two raters.

For a 4 by 4 table, these weights are as follows.

|     | $w_{11}, w_{22}, w_{33}, w_{44}$ | $w_{12}, w_{21}, w_{23}, w_{32}, w_{34}, w_{43}$ | $w_{13}, w_{31}, w_{24}, w_{42}$ | $w_{14}, w_{41}$ |
|-----|---|---|---|---|
| (a) | 1 | 0.89 | 0.56 | 0 |
| (b) | 1 | 0.67 | 0.33 | 0 |

Table 6.12 is a good candidate for the application of weighted kappa. It shows the relation between teacher assessments of spelling for pupils at the end of Year two (which is the end of Key Stage One in England and Wales) and their scores on the corresponding standard assessment task (SAT). In other words, we have two ratings of the same underlying variable, spelling ability. The data were collected as part of the evaluation of the 1991 pilot of national assessment, following the introduction of the National Curriculum, and they are described in more detail in Plewis (1997). Each pupil was allocated to a level which varied from 0 ('working towards' level one) up to three, with the expected level for pupils of this age being two. It was a subset of these data which was used in Section 6.5. However, we ignore variability between schools here.

We see that the SAT rating was greater than the teacher assessment about five times as often as the teacher assessment was higher than the SAT but the disagreements were only more than one level in 0.7% of cases. With the first set

**Table 6.12** Agreement between SAT and teacher assessment: spelling

| Teacher assessment | Standard assessment task |       |       |       |            |
|---|---|---|---|---|---|
|   | 0 | 1 | 2 | 3 | Total |
| 0 | 72 | 105 | 8 | 0 | 185 (3%) |
| 1 | 28 | 1411 | 859 | 42 | 2340 (34%) |
| 2 | 0 | 205 | 3057 | 693 | 3955 (58%) |
| 3 | 0 | 0 | 51 | 345 | 396 (6%) |
| Total | 100 (1%) | 1721 (25%) | 3975 (58%) | 1080 (16%) | 6876 (100%) |

of weights (a) given above, $\kappa_w=0.66$, with the second set (b) the estimate is 0.57. Because the sample is large, the standard errors are small and so we can conclude that teacher assessments of spelling agree moderately well with the levels attained in the corresponding SAT. If we ignore the weights, the estimate of kappa is 0.51 with a standard error of 0.01, a level of agreement which is somewhat lower than the weighted estimates.

However, if we continue in the spirit of this chapter, then we would like to find a way of modelling agreement rather than just describing it. This would be particularly useful if we wanted to see whether, for example, there is a difference between boys and girls in the level of agreement between teacher assessment (TA) and SATs. Following Agresti (1988), we can write a basic model for agreement as a special kind of log-linear model as introduced in Section 6.3.

Our log-linear model now is:

$$\log N_{ij}=a+r_i^{TA}+c_j^{SAT}+b(i.j)+d(i,j) \quad (6.19)$$

where $i$ ($i=0,1...3$) represents TA level; $j$ ($j=0,1...3$) represents SAT level; $N_{ij}$ is the expected cell size for cell $ij$ in Table 6.12; $r_i^{TA}$ is the main effect for TA and $c_j^{SAT}$ is the main effect for SAT.

Rather than represent the interaction between TA and SAT by all nine parameters, we estimate just two parameters. The first parameter, $b$, is the coefficient for the linear by linear association between TA and SAT (the term used in Section 6.3), where both TA and SAT are represented by the integers from 0 to 3 (or from 1 to 4). The second parameter, $d$, is the coefficient for a variable which takes the value 1 for each of the diagonal cells and 0 elsewhere. In other words, $d(i,j)=1$ for $i=j$ and 0 for $i\neq j$. Therefore, $d$ represents the extra agreement to be found in the table after allowing for the linear by linear association.

This model fits Table 6.12 well; the deviance is 8.31 with seven degrees of freedom ($p>0.3$). The estimate of $b$ is 2.2 with a s.e. of 0.10, the estimate of $d$ is 0.53 with a s.e. of 0.06. In other words, both the linear by linear association and the extra diagonal agreement are substantial.

We now want to extend the model to see if these two parameters – $b$ and $d$ – vary by sex. Table 6.13 gives the data (Plewis, 1997). The top entry in each cell is

**Table 6.13** Agreement between SAT and teacher assessment by sex

| Teacher assessment | Standard assessment task |  |  |  | Total |
|---|---|---|---|---|---|
|  | 0 | 1 | 2 | 3 |  |
| 0 | 38 | 71 | 4 | 0 | 113 (3%) |
|  | 34 | 34 | 4 | 0 | 72 (2%) |
| 1 | 14 | 884 | 483 | 18 | 1399 (39%) |
|  | 14 | 527 | 376 | 24 | 941 (28%) |
| 2 | 0 | 122 | 1460 | 283 | 1865 (53%) |
|  | 0 | 83 | 1597 | 410 | 2090 (63%) |
| 3 | 0 | 0 | 22 | 146 | 168 (5%) |
|  | 0 | 0 | 29 | 199 | 228 (7%) |
| Total | 52 | 1077 | 1969 | 447 | 3545 |
|  | (1%) | (30%) | (56%) | (13%) | (100%) |
|  | 48 | 644 | 2006 | 633 | 3331 |
|  | (1%) | (19%) | (60%) | (19%) | (100%) |

for boys, the bottom for girls. We see that girls get higher TA and SAT ratings than boys for spelling.

The log-linear model now is:

$$\log N_{ijk} = a + r_i^{TA} + c_j^{SAT} + l_k^S + A_{ik}^{TA,S} + A_{jk}^{SAT,S} + b_k(i,j) + d_k(i,j) \qquad (6.20)$$

where $k$ represents sex ($k=1,2$), and $A_{ik}$ and $A_{jk}$ are interactions between TA and sex and between SAT and sex, which must be in the model to fix the margins. We want to know whether $b_k = b$ and $d_k = d$; in other words, whether the two agreement parameters vary by sex.

First, we fit the simpler model, assuming that $b$ and $d$ do not vary by sex. This fits well with a deviance of 12.3 with 16 df. Adding the two extra parameters, which allow for variation by sex reduces the deviance by only 1.69 (two df). So we can conclude that the agreement structure of Table 6.13 is the same for boys and girls. (The weighted kappa values, using the first set of weights, are 0.66 for boys and 0.63 for girls.)

## 6.8 Concluding remarks

This chapter has introduced different approaches to the analysis of unordered and ordered categorical responses. As always, differences in approach reflect differences in the questions posed. Thus, log-linear models focus on associations between variables and can be extended to model agreement as well. Proportional odds and continuation odds models for ordered responses are analogous to multiple regression models for continuous responses, and can be adapted for use with longitudinal data. We have not considered model checking in much detail in this chapter but the ideas discussed in Chapter 5, and applied to binary responses, can be used in a similar way with categorical responses.

## Appendix 6.1

### SAS code

There are some points to note about the use of the CATMOD procedure. First, the response is defined as the inverse of the usual cumulative logits, in other words as $\log[(1 - CuP_{ik})/CuP_{ik}]$. However, we can get back to the more usual representation just by reversing the signs of all the coefficients. Second, rather than using the constraint of putting the estimate of the last category equal to zero, the effect of the last category is instead equal to minus the sum of all the other category effects.

Note that the text in italics is comment and so should not be typed.

*(A) SAS code for proportional odds analysis of Table 6.5*
data expect;
input sex ter84 count; (*i.e. Table 6.5*)
cards;
1 1 56
1 2 69

```
1 3 64
2 1 28
2 2 62
2 3 75
;
proc catmod data=expect;
weight count;
response clogits;
(i.e. cumulative logits)
model ter84=_response_ sex/predict
title='teacher expectations, year 2, reading';
run;
```

*(B) SAS code for continuation odds analysis of Table 6.6*
```
data sat1;
input sex sat count; (i.e. Table 6.6)
cards;
1 1 20
1 2 32
1 3 6
2 1 5
2 2 39
2 3 9
;
proc catmod data=sat1;
weight count;
response −1 1 0 0,0 0 1 −1 log 1 0 0,0 1 1,0 0 1,0 1 0;
```
*(this sets up the transformations needed to define the response as log continuation odds)*
```
model sat=_response_ sex/predict;
run;
```

*(C) SAS code for continuation odds analysis including school*
```
data sat2;
input sex school sat count;
cards;
1 1 1 5
1 1 2 9
1 1 3 1
2 1 1 3
2 1 2 14
2 1 3 1
1 2 1 4
1 2 2 15
1 2 3 4
2 2 1 1
2 2 2 15
2 2 3 3
1 3 1 11
```

1 3 2 8
1 3 3 1
2 3 1 1
2 3 2 10
2 3 3 5
;
proc catmod data=sat2;
weight count;
response –1 1 0 0, 0 0 1 –1 log 1 0 0, 0 1 1, 0 0 1, 0 1 0;
*(this sets up the transformations needed to define the response as log continuation odds)*
model sat=_response_ sex school/predict;
run;

*(D) SAS code for marginal homogeneity analysis of Table 6.8*
data expy12a1;
input exp1 exp2 count; *(i.e. Table 6.8)*
cards;
1 1 58
1 2 37
1 3 6
2 1 25
2 2 63
2 3 50
3 1 1
3 2 29
3 3 84
;
proc catmod data=expy12a1;
weight count;
response marginals;
model exp1*exp2=_response_;
repeated year 2;
*(this defines the response – the marginal probabilities – as a repeated measure)*
run;

*(E) SAS code for modelling unconditional change (Table 6.9)*
data expy12a2;
input exp1 exp2 sex count; *(i.e. Table 6.9)*
cards;
1 1 1 39
1 2 1 28
1 3 1 0
2 1 1 16
2 2 1 25
2 3 1 24
3 1 1 1
3 2 1 16
3 3 1 40

1 1 2 19
1 2 2 9
1 3 2 6
2 1 2 9
2 2 2 38
2 3 2 26
3 1 2 0
3 2 2 13
3 3 2 44
;
proc catmod data=expy12a2;
weight count;
response −1 1 0 0 0 0 0 0,0 0 −1 1 0 0 0 0,0 0 0 0 −1 1 0 0,0 0 0 0 0 0 −1 1
log 0 0 0 1 0 0,0 0 0 0 1 1,0 0 0 1 1 0, 0 0 0 0 0 1,1 0 0 0 0 0,0 1 1 0 0 0,1 1 0 0 0 0,0 0
1 0 0 0
*1 1 1 0 0 0 0 0 0 0,0 0 0 1 1 1 0 0 0,0 0 0 0 0 0 1 1 1,1 0 0 1 0 0 1 0 0,
0 1 0 0 1 0 0 1 0,0 0 1 0 0 1 0 0 1;
(*this defines the cumulative logits for the marginal probabilities*)
model exp1*exp2 = _response_ sex;
repeated year 2;
run;

## Exercises

Answers to those exercises marked '#' can be found at the end of the book.

1  Assume that the observed counts for the diagonal cells in Table 6.1 are equal to the expected counts. One way of doing this is to give the diagonal cells a weight of zero. Such cells are known as *structural zeros*. One reason for doing this is that some teachers might automatically give the same ratings for reading and mathematics. Is there any association between the two sets of expectations, having taken the diagonal cells out of the analysis? This is a quasi-independence model; leaving aside the diagonal cells, is a pupil more likely to get a higher rating for reading if they get a higher rating for mathematics etc?
(#)

2  Start by fitting a saturated model to Table 6.2 and then leave out terms until you arrive at a model which gives a satisfactory fit.

3  Calculate the logged continuation odds for Table 6.6, and the cumulative logits for Table 6.7.
(#)

4  Obtain a test statistic for the hypothesis of marginal homogeneity applied to Table 6.12.

5  Take teacher assessment (TA) as the response in Table 6.13 and then fit a continuation odds model, with SAT and sex as explanatory variables. Explain your results carefully, referring to Plewis (1997) as appropriate.

# 7
# Education as a Career: Event History Analysis

## 7.1 Introduction

The concept of 'career' has become increasingly used by social scientists, not only in discussions of individuals' labour market experiences, but also when thinking about marriage and housing, when *durations* in particular states (cohabitation, divorce etc) and *transitions* between different states (for example, renting to owner occupation) are of particular interest. In education, the idea of a career is especially germane to children's experiences before they reach the age of compulsory schooling, because of the variety of services available to parents of young children in many Western societies. A young child might have several episodes at home cared for by one or other parent, punctuated by times with a childminder, attendance at playgroup, and a period at nursery school. Understanding why there is variation in preschool careers across children could be important both from a developmental perspective and for policy. Career is also important after the end of compulsory schooling when young adults make choices about pathways through further and higher education, a process which can last for many years. Even during the years of compulsory schooling, although some transitions are fixed for everyone, such as transfer to secondary school, other changes, for example moves from one secondary school to another, can vary from pupil to pupil.

A proper understanding of how the educational system is functioning requires data on educational careers as well as the more usual official statistics on pupil attainments, pupil–teacher ratios, numbers and ages of teachers and so on. However, the production of official education statistics, at least in the United Kingdom, is not geared to providing the kinds of longitudinal data implied by the idea of educational careers. Thus, we must rely on the comparatively few research studies which have generated this kind of data. Here, we use data on preschool experiences from a relatively small research study, to show how the methods of analysis described in this chapter can further our understanding about careers. We should note that event history data are usually collected in longitudinal studies, although they can be obtained completely retrospectively.

114   *Education as a Career*

In the next section, different kinds of event history data are described. The fundamental concepts are then introduced, followed by a description of graphical methods for exploring event history data. In Sections 7.5, 7.6 and 7.7, some simple statistical models are discussed, which are then applied to the preschool histories mentioned above.

## 7.2   Types of event history data

Statistical methods for the analysis of data on careers are often referred to as *event history analysis*, sometimes as models for *stochastic processes* and sometimes as *Markov models*. Event history analysis is used in this chapter. The methods have evolved from methods developed in epidemiology, demography and in reliability testing in industry. In each of these contexts, the terms *survival analysis* and *hazard models* are widely used. Some of the vocabulary used is, however, confusing and so a dictionary of commonly used terms is given in Appendix 7.1.

Let us start with the simplest kind of event history, when there is just a single unrepeated event (or change of state or transition) and hence just one episode. Our interest is in explaining the variability of the duration of this episode. Although simple, it is data of this kind from clinical trials in medicine, life tables in demography and product testing in industry that have exerted a powerful influence on methods for the analysis of event histories. An example of a single unrepeated event in education is the first occasion at which a child goes to a day care or educational institution, and so the episode starts at birth and ends at the age at which the child is first cared for outside the home for some part of the day. This is illustrated in Figure 7.1 and we will be interested in why the episodes vary in length or duration – from four months for child A to 60 months for child C.

A rather more complicated event history is generated when we have single

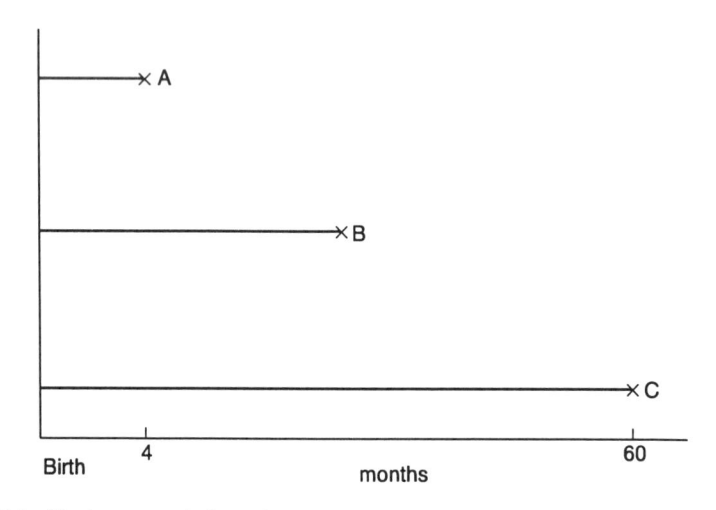

**Figure 7.1**   Single unrepeated events

repeated events. To continue with the preschool theme, we might observe a series of episodes before a child starts compulsory schooling. However, we do not distinguish between the different kinds of preschool facilities the child might attend. We might choose to analyse the first episodes separately from the second and subsequent episodes. However, if we choose to analyse all episodes together, we will need to recognize that episode durations will not, in general, be independent, because some children will have shorter durations, on average, than others.

Often, we will want to distinguish between different kinds of preschool experiences, and we then enter the realm of *competing risks* models. Competing risks occur when two or more events can determine the end of an episode. For example, we might be interested in the time which elapses before a child first attends a group environment such as a nursery or a playgroup. However, for some children, their first episode away from home could be with a childminder and thus transitions to childminders and to group care are competing risks.

If we put together repeated episodes and competing risks, we generate a full event history covering, in our case, the period from birth to the start of compulsory schooling. This is illustrated in Figure 7.2, and we might be interested both in episode durations and influences on transitions from one type of episode to another.

Before moving on to describe how event history data might be analysed, we meet the dataset to be used throughout this chapter (dataset 7.1 on the disk). The initial sample consisted of 255 women and their first born children living in the greater London area and born in the British Isles or Eire. The sample was selected on the basis of the employment intentions of the women. One group intended not to return to work following the birth of their child. The other three groups consisted of women intending to return to full-time work before their child was nine months old, classified according to the type of care arrangement they had chosen for the child: relative, childminder or nursery. The women were interviewed on five occasions, when the children were around 4, 11, 18, 36 and 72 months old. Attrition was low; by the time the children were six years old, the sample had been reduced by only 10% to 230. Further details of the study can be found in Brannen and Moss (1991).

The definition of what constitutes an episode is not always clear-cut in this kind of work. Here, an episode is defined as:

a spell of 'care' either with a parent, or in another setting, the end of which was marked by a definite change in the care environment.

The first episode started at birth and was always at home, the second episode

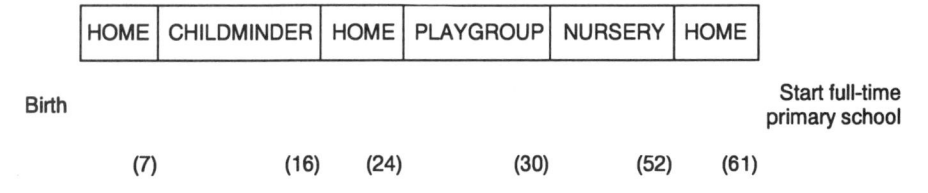

| HOME | CHILDMINDER | HOME | PLAYGROUP | NURSERY | HOME |
|------|-------------|------|-----------|---------|------|

Birth                                                                                     Start full-time
                                                                                          primary school

        (7)            (16)         (24)          (30)         (52)        (61)

**Figure 7.2**  Hypothetical preschool history

116  *Education as a Career*

was therefore always away from home, but there were no restrictions on subsequent episodes. Altogether, 1181 episodes were coded for 249 children, with a range from one to 12 episodes per child. We are interested in the variability of the preschool histories of the sample children, and whether we can account for this variability by explanatory variables defined at the child or mother level, such as the mother's educational qualifications and the child's sex, and by explanatory variables defined at the episode level such as type of care arrangement. The references to 'levels' here suggests that the multilevel techniques described in Chapters 3 and 4 might also have a part to play in the analysis of event history data (see Section 7.7).

## 7.3 Fundamental concepts in event history analysis

The easiest way of presenting the fundamental concepts in event history analysis is to do so in the context of a single unrepeated event. For our example, this means the episode starting at birth and ending with a transition to any form of care outside the home. Also, we assume to start with that there was no attrition from the sample.

We start with the notion of a *risk set*. This is the set of children at time $t$ (and thus who are aged $t$ months) who have yet to experience the event of a move to outside care. The size of the risk set inevitably decreases with time, sometimes quickly sometimes slowly, and, in principle, could eventually become zero. This would happen with these data if we regard the start of compulsory schooling as a transition into outside care. Being aware of what the risk set is, not only in event histories but also in cross-sectional studies, can help us to avoid making careless inferences. For example, a sample of two year olds not attending preschool is different from a sample of three year olds not attending preschool, not only because of the age difference but also because of any systematic differences, perhaps in home circumstances, which lead some children to start preschool earlier than others. In other words, the risk set at two years is different from (and larger than) the risk set at three years.

The decline in the size of the risk set over time (or age) can be represented by the *survivor function*, or the *survival curve*. We met survival probabilities in Chapter 6 (row five of Table 6.5). The survival probability is the complement of (i.e. one minus) the cumulative probability. The survivor function describes how the survival probabilities change with time; it starts at one and can never increase with time, ending up at zero if the risk set becomes empty at some time. Demographers represent it by the life table. Figure 7.3 illustrates the idea: there are two survival curves corresponding to two groups, and one curve declines more sharply than the other so that the median survival time for group A is less than the median survival time for group B. If we call the survivor function $S(t)$ then:

$$S(t) = 1 - F(t) = \text{Prob}(T > t) \tag{7.1}$$

where $t$ is time, $T$ is the time of the event, sometimes called the 'failure' time, and $F(t)$ is the cumulative probability, or distribution function (row four of Table 6.4).

Also:

$$S(t_1) \geq S(t_2) \quad \text{for all } t_2 > t_1 \tag{7.2}$$

## 7.3 Fundamental concepts in event history analysis

**Figure 7.3** Two survival curves

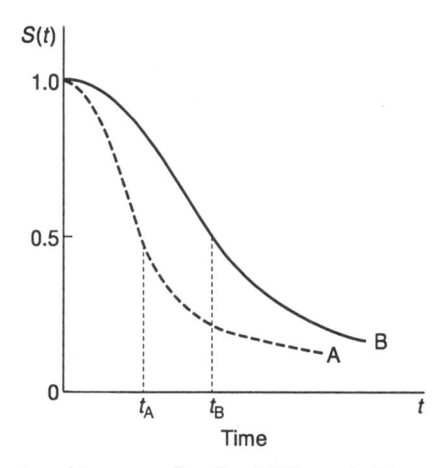

$t_A$ and $t_B$ are median (i.e. 50%) survival times

In other words, we rule out the possibility of anyone moving back into the risk set once they have left it.

Another key concept in event history analysis, especially when we want to model durations, is the *hazard*. It can be thought of as the rate at which events occur, or as the risk of an event occurring at a particular time $t$. In essence, it is the conditional probability of an event happening at time $t$, given that it has not happened before $t$. If we call the hazard function $h(t)$ and we call the probability density function (pdf) for the duration $f(t)$, then:

$$h(t) = f(t)/S(t) \tag{7.3}$$

Unlike the survivor function, the hazard function can have any shape. It can be constant over time, it can increase or decrease steadily over time, it can both increase and decrease over time. This shape can tell us quite a lot about the underlying process. We look at representations of hazards and hazard models in more detail in Sections 7.5 and 7.6.

Sometimes it is useful to examine the *cumulative hazard function* or integrated hazard as this can give us a better idea about the shape of the hazard. There is a simple link between the cumulative hazard, $H(t)$, and the survivor function, namely:

$$H(t) = -\log S(t) \tag{7.4}$$

The last of the fundamental concepts in event history analysis introduced here is censoring. Now we drop our unrealistic assumption of no sample loss over time. There are two important kinds of censoring: right censoring and left censoring. Right censoring occurs when we know the start time of an episode but not when it ends. Left censoring occurs when we know of the existence of an episode but not when it starts. Left censoring presents a number of rather awkward problems for event history analysis which would take us beyond the scope of this chapter. It is clearly not a problem for our chosen example because all these episodes start at birth. However, right censoring is a pervasive problem with event histories, essentially because, as we discussed in Chapter 4, some degree of attrition is almost inevitable in longitudinal studies. It is not uncommon to find statistical analyses of educational data based solely on complete cases, and hence the missing data are ignored. We have already seen in Chapter 4 that this is undesirable when analysing educational growth. It is also undesirable to ignore right-censored episodes when analysing durations, because to do so introduces bias, particularly when estimating the hazard function. This bias arises because, on average, censored episodes last longer than uncensored ones, and so the estimated hazard function based only on uncensored observations will be too high.

In the next section, we describe ways of allowing for right censoring when constructing survival curves. The only assumption we do need to make – and it is quite a strong assumption – is that the process which generates the censoring is unrelated to the process which generates the event histories. Another way of putting this is to say that the censored observations should be missing at random, or ignorable, in the sense that these phrases are used in Chapter 4. In our case, this means that children are not lost from the study because they have particular kinds of day care histories.

Now that we have allowed for the possibility of right censoring, we should amend our definition of the risk set: it is the set of children at time $t$ yet to experience an event *and* whose episodes have not been censored.

### 7.4 Constructing survival curves

It is a straightforward matter to construct sample survival curves in the absence of right censoring. All we need to do is to order the survival times from lowest to highest, calculate the survival probabilities, which start at one and decline, and then plot these against time. The situation is a little more complicated with censoring and the *Kaplan–Meier* method is then often used. A detailed description of the Kaplan–Meier method can be found in Chapter 6 of Aitkin *et al.* (1989), but, in essence, the method works by ordering the survival times ($t_j$), calculating the risk set ($r_j$) at each ordered survival time (including any observations which are censored at precisely that time), noting the number who do not survive at each survival time ($d_j$), calculating the proportion who survive ($1-(d_j/r_j)$) and then taking the product of the survivor proportions up to each survival time.

We illustrate the Kaplan–Meier method for the first preschool episodes in our example. These are the second episodes as the first episode is always at home. Most of these 234 episodes fell into four types – relative, childminder, nursery and playgroup – and only 2.1% of them were censored. The estimated survival curves for the four types were obtained using the Kaplan–Meier procedure in the SURVIVAL menu in SPSS and are shown in Figure 7.4. We see that the curve for the nursery group is considerably different from the other three types, decreasing relatively slowly up to 24 months. The sample median survival time ($m$) is 26 months for the nursery group, compared with 12 months for the childminder and playgroup types and nine months for relatives. (The mean survival times are all a little greater than the medians. It is very common to find skewed distributions of survival times, with a long tail on the right towards longer

## 7.4 Constructing survival curves

times.) The 95% confidence intervals for the median survival times ($M$) for the four types are shown in Table 7.1. We see that the interval for the nursery group does not overlap the confidence intervals for the other three types.

Having obtained the survival curves, it is straightforward to produce the cumulative hazards, $H(t)$, and these are shown in Figure 7.5. We see that for

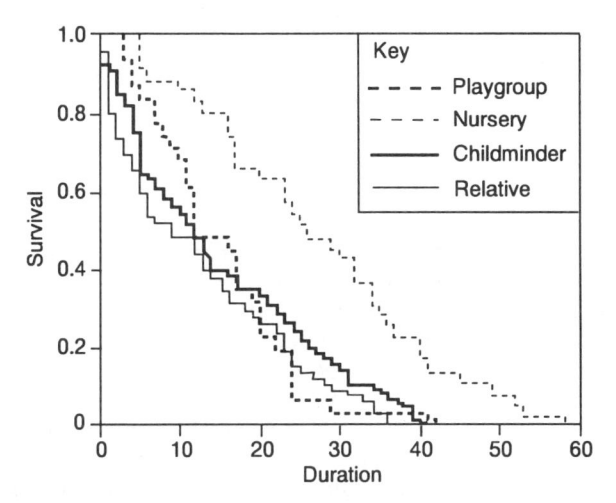

**Figure 7.4** Survivor functions by type of preschool

**Table 7.1** Median survival times by group

| Preschool type | Sample size | Number censored | Median (m) | 95% confidence limits for M |
|---|---|---|---|---|
| Relative | 65 | 1 | 9 | 3.56; 14.44 |
| Childminder | 102 | 3 | 12 | 9.01; 14.99 |
| Nursery | 35 | 0 | 26 | 17.89; 34.11 |
| Playgroup | 32 | 1 | 12 | 7.91; 16.09 |

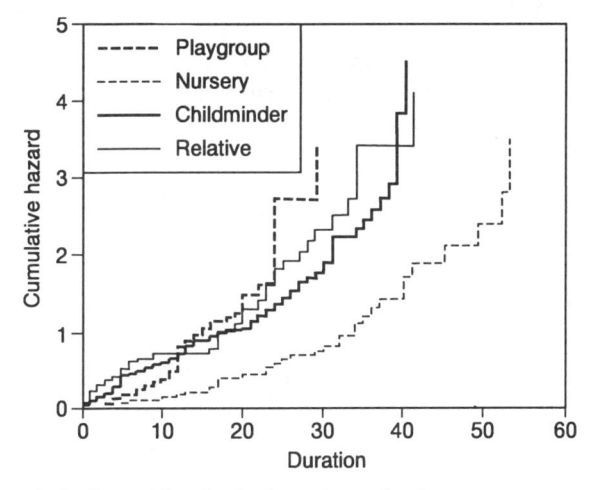

**Figure 7.5** Cumulative hazard function by type of preschool

120  *Education as a Career*

each of the four groups, the underlying sample cumulative hazards curve upwards rather than being straight lines. This suggests that the population hazard increases more and more rapidly as the duration increases. As we shall see in the next section, this tells us something about the hazard distribution.

Examining survival curves and cumulative hazards is an important first exploratory step in any event history analysis. However, in keeping with the spirit of this book, we are likely to understand the underlying process better if, having explored our data, we go on to try to construct a realistic statistical model, and this is the subject of the next section.

## 7.5 Statistical models for event history data in continuous time

Some of the steps we have to take when we construct statistical models are, as we have seen from earlier chapters, common to all types of data. We have to distinguish between the response and the explanatory variables, we have to decide how to link the explanatory variables to the response and so on. However, event history data do present us with some special choices.

First, we have to decide just what our response is. We can model the episode duration directly or we can model the hazard function. Developments in statistical theory for event history data have tended to focus on hazard models, especially *proportional hazards* models. One of the reasons why hazard models are preferred is that they handle censored data easily within the framework of likelihood based estimation. However, if there is little censoring, then there are advantages in using the duration time, or the log of duration, as the response and then applying standard least squares regression methods as described is Chapter 2. Models for duration are often known as *accelerated life* models. Both kinds of model are illustrated below.

The second choice we have is to decide what distributional assumptions we should make, either for the hazard or for the duration, or indeed whether any distributional assumption is necessary. The simplest assumption we can make is that the durations follow an exponential distribution which implies that the hazard is constant over time (and therefore the cumulative hazard is a straight line rather than the curves of Figure 7.5). In other words, exponentially distributed durations imply that the risk of an event or a transition does not vary with time or age. However, it is more common for hazards to exhibit *duration dependence* of some kind, either to increase or decrease steadily over time or even to change direction over the time span of interest. For example, we might expect a preschool episode to be most at risk of ending early in the episode as parents quickly discover aspects of the arrangement they do not like; the hazard then would start high and later decrease. If we think about the hazard of leaving a one-year postgraduate course, then this might be low to start with but might increase as the volume of work mounts and examinations loom. Hence, we really need a more flexible distribution than the exponential to describe hazards, which is why statisticians often use the Weibull distribution for durations. The Weibull includes the exponential as a special case, but also allows hazards either to increase or decrease (but not to change direction). The hazard for the Weibull distribution takes the form:

$$h(t) = \lambda \alpha t^{\alpha-1} \tag{7.5}$$

## 7.5 Statistical models for event history data in continuous time

where $\alpha$ is the parameter of interest and $\lambda$ is just a scaling parameter. The Weibull survivor function is:

$$S(t) = \exp(-\lambda t^\alpha) \quad (7.6)$$

Examples of Weibull hazards are shown in Figure 7.6. When $\alpha = 1$ the Weibull reduces to the exponential or constant hazard, with $h(t) = \lambda$ and the cumulative hazard, $H(t) = \lambda t$. The cumulative hazard is then a straight line through the origin with slope equal to the exponential parameter $\lambda$.

Sometimes we want to know the shape of the hazard in order to understand the process better. But, quite often, our main interest is in the effects of explanatory variables on the hazard, rather than the shape of the hazard function itself. It is for such situations that the *Cox proportional hazards* model is especially useful, because it makes no assumptions about the shape of the hazard. The Cox model is defined and applied below.

Our third main choice is about how we treat time. Do we work with continuous time models or with discrete time models? Our decision will essentially be determined by how we collect our data. If we are able to date all the events to a high degree of accuracy – say to the nearest day – then continuous time models will be appropriate. But if, as is often the case with educational data, our dating is less fine-grained, then it could be better to treat time as discrete. In our example, we dated preschool transitions in terms of completed months of age, which leads to some problems with episodes which lasted less than a month. This dating also created a lot of ties, which are not strictly permissible within a continuous time model. As we shall see, discrete time models are closely related to the models for binary data described in Chapter 5.

We now see how some models work in practice by applying a sequence of them to the first preschool episodes for which the survival curves were estimated in Section 7.4. Throughout, we consider three explanatory variables: the four types of preschool as before, but here represented as three dummy variables; the

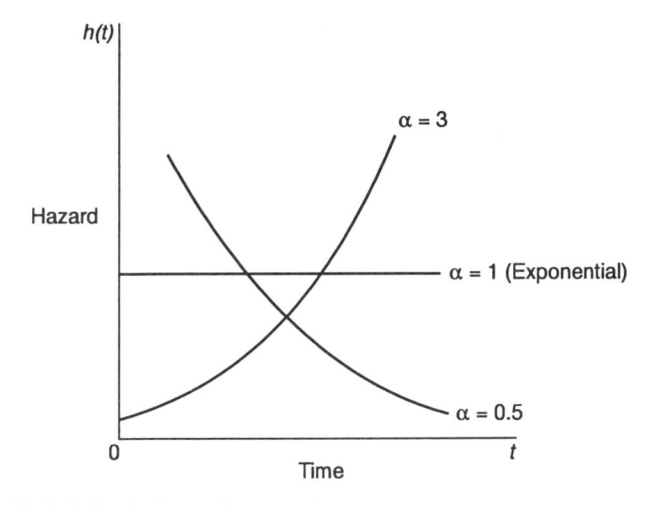

**Figure 7.6** Weibull distributions with parameter $\alpha$

## 122  Education as a Career

child's age at the start of the episode; and the mother's educational level varying on a scale from zero (no qualifications) to six (degree). (Throughout this chapter, we consider only explanatory variables which do not vary with time.)

We start with a model for log duration. We might expect this model to work quite well with these data because there is so little censoring. Rather than omitting the censored cases, we treat the episode length as the duration up to the point of censoring. (But we should not do this if there were a more substantial amount of censoring, say greater than 10%.) Let us start with a model with just one explanatory variable, mother's educational level. Our model is:

$$\log_e t_i = a + bx_i + e_i \tag{7.7}$$

where $t_i$ is duration for individual $i$, and $x$ is mother's educational level.

This is a simple regression model with a residual $e_i$, familiar to us from Chapter 2, which we can estimate easily using least squares. There is an implied assumption that $\log t$ is Normally distributed given $x$, equivalent to saying that the durations follow a log normal distribution given $x$. How reasonable is this? We can use Q-Q plots of the studentized residuals to form a judgement about this, just as we did in Chapter 2. Figure 7.7 indicates that there are some, possibly not too severe, departures from Normality.

The estimate of $b$ is 0.082 with a standard error of 0.043. This means that a unit increase in the scale for mother's education leads, on average, to an increase of 0.082 months in log duration. In other words, $\log t_2 - \log t_1 = \hat{b}$ for all differences of one scale unit. This is equivalent to $t_2 = kt_1$ where $k = \exp(0.082) = 1.09$, and so the durations are accelerated by a factor $k$ if $\hat{b} > 0$, as here, decelerated by $k$ if $\hat{b} < 0$ (and are, of course, unaffected by $x$ if $\hat{b} = 0$). The estimate of $a$ is 1.95

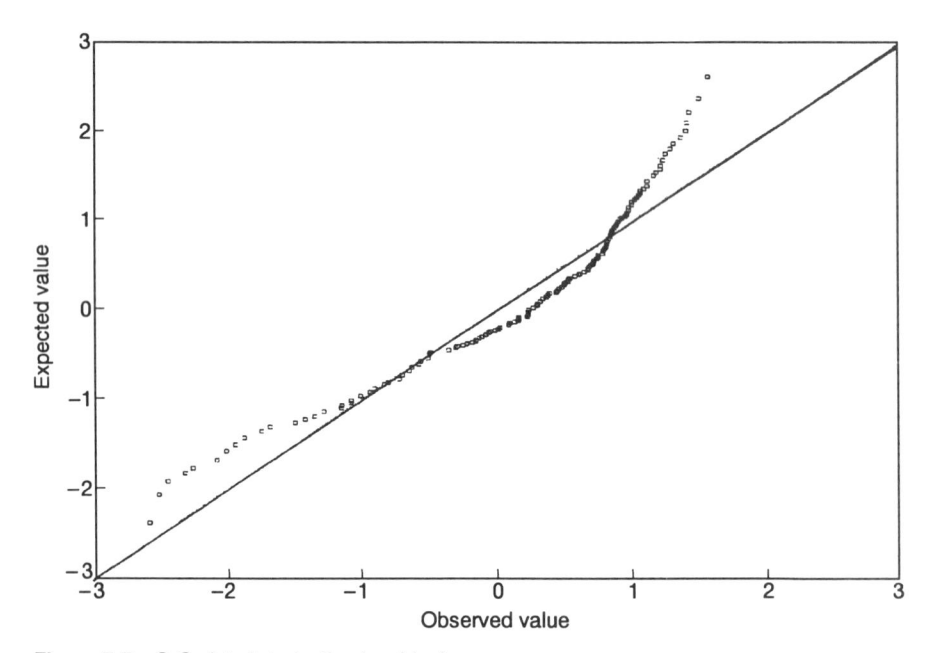

**Figure 7.7**  Q-Q plot of studentized residuals

## 7.5 Statistical models for event history data in continuous time

with a s.e. of 0.18, showing that the expected duration for the first day care episode for a mother with no educational qualifications is exp(1.95)=7 months.

We now extend the model to include all the selected explanatory variables, just as one might go from a simple to a multiple regression model. Our model now is:

$$\log t_i = a + b_1 x_1 + b_2 x_2 + b_3 d_1 + b_4 d_2 + b_5 d_3 + e_i \qquad (7.8)$$

where $x_1$ is start age, $x_2$ is mother's education and $d_1$, $d_2$ and $d_3$ are dummy variables for preschool type. The estimates, their standard errors and 95% confidence intervals are given in Table 7.2. We see that the picture is rather different from the one obtained from the comparison of the survival curves in Figure 7.4. There, the nursery group had longer durations on average. Now we see, after controlling for age at the start of the first preschool episode, that relative and childminder episodes are shorter compared with playgroups, but nursery episodes are not. However, durations become shorter as age at the start rises. This model accounts for 15% of the variance in durations, with mother's educational level no longer an important explanatory variable after the other variables have been included. When we plot the studentized residuals against start age, we get Figure 7.8. This plot suggests there are two groups within the data – those with an early start (less than nine months) and those with a first episode starting after 24 months. We know that the sample consisted of returners to full-time work and a smaller group of non-returners, and a more detailed analysis might wish to incorporate this aspect of the design into the analysis. Moreover, we know that playgroup episodes cannot usually start before a child is two years old and so, in a fuller analysis, we would perhaps explore possible interactions between preschool type and start age (see Exercise 2 at the end of this chapter). However, for now, we move on to a model for the hazard.

We can write the Cox proportional hazards (PH) model for one explanatory variable as:

$$h(t) = h_0(t) \exp(bx) \qquad (7.9)$$

where $h(t)$ is the hazard which may or may not be duration dependent (which is equivalent to saying that the durations may or may not be exponentially distributed), and $h_0(t)$ is an unknown baseline hazard. In other words, we do not need to know the shape of this baseline hazard, which could be constant, monotonically increasing or decreasing, or even have a 'bathtub' or an 'inverted bathtub' shape. What we consider as the baseline will be determined by which explanatory variables are in the model.

**Table 7.2** Log duration model: estimates

| Explanatory variable | Estimated regression coefficient | Standard error | 95% confidence limits for $b_i$ |
|---|---|---|---|
| Constant | 3.86 | 0.55 | 2.78; 4.93 |
| Start age | −0.043 | 0.014 | −0.071; −0.015 |
| Mother's education | 0.02 | 0.043 | −0.065; 0.106 |
| Relative* | −1.78 | 0.46 | −2.69; −0.87 |
| Childminder* | −1.49 | 0.46 | −2.39; −0.59 |
| Nursery* | −0.63 | 0.50 | −1.62; 0.37 |

*compared with playgroup as baseline category.

## 124  *Education as a Career*

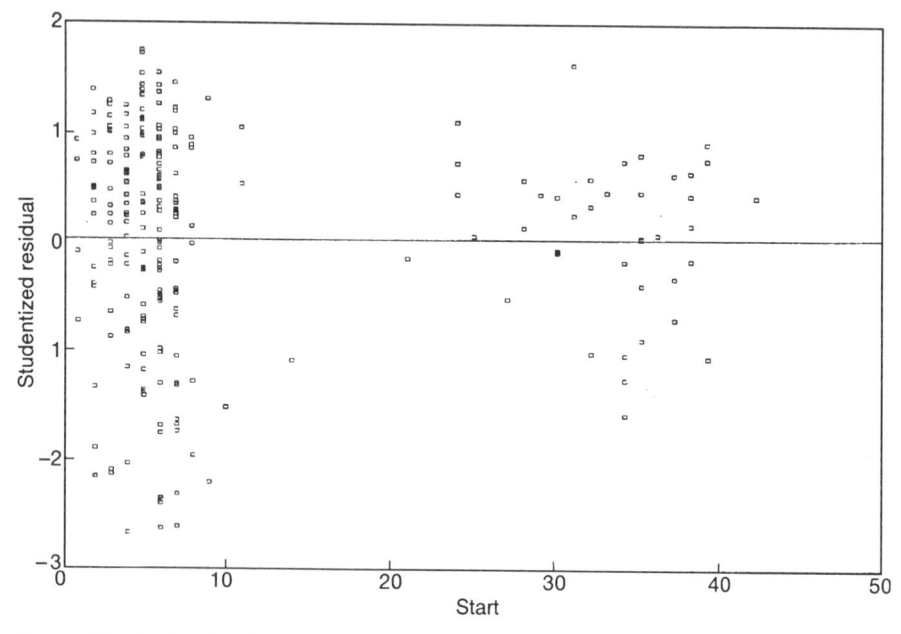

**Figure 7.8**  Scatterplot of studentized residuals by episode start

We can transform this multiplicative model to an additive one, just as we did in Chapter 6, by taking logs so that:

$$\log h(t) = a(t) + bx \tag{7.10}$$

where $a(t)$ is the log of the baseline hazard at time $t$. This is another kind of generalized linear model which requires a special method of estimation (see, for example, Cox and Oakes, 1984).

We again use mother's educational level as our explanatory variable, and focus on the first preschool episode. Our baseline is mothers without any educational qualifications. The estimated value of $b$ is –0.04 with a standard error of 0.04 (obtained from the Cox regression procedure within the SURVIVAL menu in SPSS). We would expect to find the sign of the coefficient in a hazard model to be the opposite of that in a model for duration because the greater the hazard the shorter the duration. The coefficient is smaller here relative to its standard error than it was in the model with log duration as the response. For every increase of one unit on the scale for mother's education, the ratio of the hazards is $\exp(-0.04) = 0.96$, hence the term 'proportional hazards'. We must recognize, however, that the proportionality is an assumption that is built into the way the model is specified. It implies that the ratio of the hazards does not vary over time. There are ways of testing this assumption, for example by including an interaction between the explanatory variable and time in the model, but they take us beyond the scope of this chapter. The proportionality assumption has proved to be a robust one for many datasets, but it should not be taken for granted. (Any more than it should with the proportional odds models in Chapter 6.)

We can easily extend the Cox model to include several covariates, just as we extended the accelerated life model. The corresponding results are given in Table 7.3. We find a good deal of consistency between the inferences from the Cox proportional hazards model, which properly accounts for censoring, and the model for log duration, which does not. The only slight difference between the two models is in the sign of the estimate for mother's education, although both coefficients are small.

The Cox model tells us nothing about the shape of the hazard, which, in some circumstances, is a disadvantage. Therefore, we would like to try to find a model which is a little more informative about the hazard, and which we can estimate reasonably easily. Hence, we make the assumption that the durations follow a Weibull distribution; an equivalent assumption is that the log durations have an extreme value (or Gumbel) distribution. We cannot use SPSS to estimate a Weibull model but we can make use of the WEIBULL macro for a hazard model in GLIM4 (Francis *et al.*, 1993), which provides estimates both for a Weibull model and for the special case of the exponential distribution. (The GLIM code is given in Appendix 7.2.) The estimates are given in Table 7.4; first for the exponential and then for the Weibull. The differences between the two models, and between these models and the Cox model in Table 7.3, are not marked and each model tells essentially the same story – shorter durations for relative and childminder episodes compared with playgroups, shorter durations the later the age at start, and no effect of mother's education.

We can compare the fit of the Weibull and exponential models and we find the Weibull is significantly better – the difference in deviances is 14.8 with one df ($p<0.001$). The extra shape parameter for the Weibull (i.e. $\alpha$) is estimated to be

**Table 7.3** Cox proportional hazards model: estimates

| Explanatory variable | Estimated regression coefficient | Standard error | 95% confidence limits for exp(b) |
|---|---|---|---|
| Start age | 0.056 | 0.013 | 1.03; 1.08 |
| Mother's education | 0.04 | 0.041 | 0.96; 1.12 |
| Relative* | 1.61 | 0.41 | 2.25; 11.2 |
| Childminder* | 1.36 | 0.40 | 1.77; 8.5 |
| Nursery* | 0.52 | 0.45 | 0.70; 4.04 |

*compared with playgroup as baseline category.

**Table 7.4** Exponential and Weibull models: estimates

| | Exponential | | Weibull | |
|---|---|---|---|---|
| Explanatory variable | Estimated regression coefficient | Standard error | Estimated regression coefficient | Standard error |
| Constant | −4.3 | 0.48 | −5.35 | 0.48 |
| Start age | 0.046 | 0.012 | 0.057 | 0.012 |
| Mother's education | 0.01 | 0.04 | 0.012 | 0.04 |
| Relative* | 1.47 | 0.41 | 1.73 | 0.41 |
| Childminder* | 1.27 | 0.40 | 1.51 | 0.40 |
| Nursery* | 0.72 | 0.45 | 0.87 | 0.45 |

*compared with playgroup as baseline category.

1.24 (compared with the value of one for the exponential), suggesting that the hazard does increase with time and therefore that there is positive duration dependence. This is consistent with the graphs of the cumulative hazards presented in Figure 7.5. After fitting the Weibull model in GLIM4, we can then plot the adjusted, or variance stabilized, residuals. The Q-Q plot is shown in Figure 7.9. We see that the residuals lie more or less on a straight line, although they are a little less dispersed than expected. The Weibull assumption appears to be reasonable.

## 7.6 Statistical models for event history data in discrete time

Up to now, we have modelled our event history data in continuous time. By doing so, we have reflected the way in which many education career changes take place; they happen haphazardly and can occur at any point in time. However, although most events happen in continuous time, our methods for measuring when these events take place are less precise, especially when some degree of retrospective data collection is used. Often the best we can do is to measure to the nearest month or year, and even those measurements can be fallible. As we have seen in our example, we have dated events to the nearest month, which is reasonably fine-grained although clearly not exact. We can, however, construct discrete time models for event histories, models which may be somewhat less appropriate in terms of the way the actual process unfolds but much better suited to the data at hand. We will see that these discrete time models have a number of similarities to the models presented in the previous chapters for binary and categorical data. We should also recognize that some events do genuinely happen in discrete time; for example, pupils usually only change sets or streams in secondary school at the end of a school year.

In order to estimate a model in discrete time, we first need to divide the range of durations into intervals of equal duration. (If the intervals are not of equal duration then the analysis becomes rather more complicated.) The actual

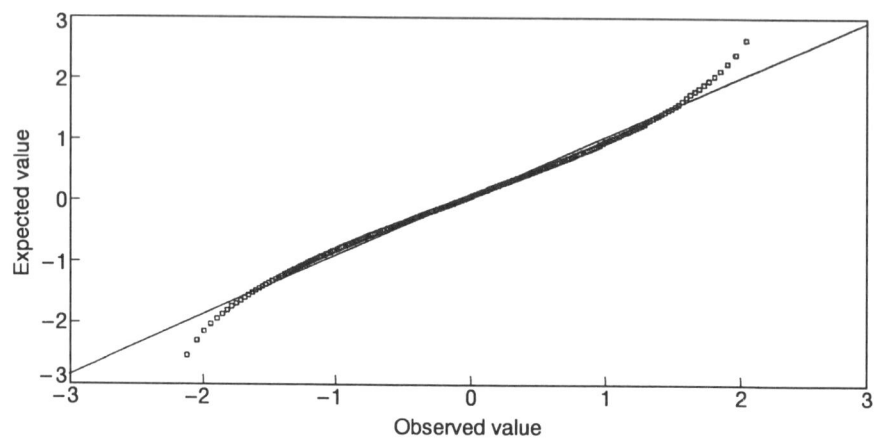

**Figure 7.9** Q-Q plot of variance standardized residuals

number of intervals will depend on the data; the more intervals the closer we come to a continuous model. On the other hand, narrow intervals might not help us to overcome imprecision in dating events. For our preschool history data, we divide the range into 20 intervals, each of three months, because the shortest duration for the first day care episode was less than one month and the longest was 58 months.

In principle, we can calculate the hazard for each of these intervals simply as the number of events ($d$) during the interval divided by the number at risk ($r$) at the beginning of the interval:

$$h_j = d_j/r_j \qquad (7.11)$$

where $j = 1, 2 ... J$ is the chosen interval.

Of course, we want to relate the discrete time hazard, $h_j$, to our chosen explanatory variables. If these are few in number, and are categorical, then we can estimate $h_j$ for each combination of these explanatory variables. In essence, we then have a table with as many cells as there are categories and intervals, and each cell contains the hazard with the numerator being $d_j$ and the denominator $r_j$. This produces a dataset which can be analysed using a model for binomial data as described in Chapter 5. An example of this method of discrete time analysis can be found in Aitkin *et al.* (1989, p.314).

Unfortunately, this approach soon breaks down as the number of explanatory variables builds up, and if any of the explanatory variables are continuous rather than categorical. What we can do instead is to create a file that is based on person intervals, sometimes called person epochs, rather than just on persons. The response takes the value zero for each person for all intervals except the interval in which the event occurs, when it is one. If the duration is right censored, then the response is always zero for all intervals up to and including the interval in which censoring takes place. The conversion from a case- or person-based file to a person interval file is illustrated in Figure 7.10.

| Case | Duration | Censor | Type | Start |
|---|---|---|---|---|
| 01 | 0 | 0 | 2 | 7 |
| 02 | 13 | 0 | 3 | 11 |
| 03 | 10 | 1 | 1 | 40 |

↓

| Case | Response | $D_1$ 0–2 | $D_2$ 3–5 | $D_3$ 6–8 | $D_4$ 9–11 | $D_5$ 12–14 | $D_6$ 15–17 | Type | Start |
|---|---|---|---|---|---|---|---|---|---|
| 01 | 1 | 1 | 0 | 0 | 0 | 0 | 0 | 2 | 7 |
| 02 | 0 | 1 | 0 | 0 | 0 | 0 | 0 | 3 | 11 |
| 02 | 0 | 0 | 1 | 0 | 0 | 0 | 0 | 3 | 11 |
| 02 | 0 | 0 | 0 | 1 | 0 | 0 | 0 | 3 | 11 |
| 02 | 0 | 0 | 0 | 0 | 1 | 0 | 0 | 3 | 11 |
| 02 | 1 | 0 | 0 | 0 | 0 | 1 | 0 | 3 | 11 |
| 03 | 0 | 1 | 0 | 0 | 0 | 0 | 0 | 1 | 40 |
| 03 | 0 | 0 | 1 | 0 | 0 | 0 | 0 | 1 | 40 |
| 03 | 0 | 0 | 0 | 1 | 0 | 0 | 0 | 1 | 40 |

**Figure 7.10** Converting duration data into person interval (person epoch) data for discrete time analysis

One rather convenient way of doing this is to use the SURV command in MLn, and the outcome of this conversion is dataset 7.2 on the disk. The six time dummies in Figure 7.10 represent the six three month intervals up to 18 months; for the full dataset there would, of course, be 20 of these dummies. It can be seen that the first person produces just one record in the person interval file because the duration is less than one month, the second produces five records with the final value of the response being one, and the third produces four records which, because the episode is censored at ten months, have all responses equal to zero.

We can treat the response in the bottom section of Figure 7.10 as a binary, or 0–1 variable and hence we can apply the method of logistic regression to it (see Chapter 5). This will give us estimates of the effects of the explanatory variables on the discrete time hazard. The theoretical justification for this can be found in, for example, Singer and Willett (1993). This model is strictly a proportional odds model rather than a proportional hazards model. However, if we use the complementary log-log function as a link, rather than the logit link, then the model is a proportional hazards model. The differences between the two models are usually small, especially if the hazard is low.

For just one explanatory variable, we write these two discrete time models as:

$$\text{logit } P_{ij} = a_j + bx_i \qquad (7.12)$$

where $P_{ij}$ is the probability of individual $i$ experiencing an event in duration interval $j$, given that they have not previously experienced the event. We can think of this as the hazard of experiencing an event in interval $j$. The $a_j$ are a series of constants, one for each interval.

With a complementary log-log link, we just have:

$$\log[-\log(1 - P_{ij})] = a_j + bx_i \qquad (7.13)$$

The estimates from the two models, obtained using GLIM4 (see Appendix 7.2), are shown in Table 7.5. One of the advantages in using GLIM4 for problems of this kind is that we can use the ELIMINATE command to save having to estimate values for each of the time dummies, which are often nuisance parameters. We see that the estimates for our explanatory variables are reasonably close, those for the model with a logit link tending to be a little higher than those for the model with a log-log link. The story they tell is the same as the one told by the estimates for the continuous time models in Tables 7.2–7.4.

**Table 7.5** Discrete time model: estimates

| Explanatory variable | Logit Estimated regression coefficient | Standard error | Log-log Estimated regression coefficient | Standard error |
|---|---|---|---|---|
| Start age | 0.069 | 0.016 | 0.058 | 0.013 |
| Mother's education | 0.038 | 0.046 | 0.036 | 0.041 |
| Relative* | 2.02 | 0.52 | 1.68 | 0.42 |
| Childminder* | 1.73 | 0.52 | 1.43 | 0.41 |
| Nursery* | 0.78 | 0.56 | 0.57 | 0.46 |

*compared with playgroup as baseline category.

Although the estimates for the time dummies are often not of great interest, just as the distribution of the baseline hazard is not important in a Cox model, nevertheless we might sometimes like to get some idea of the shape of the hazard function by examining them. They are shown in Table 7.6 for the two link functions; in both cases the hazard stays fairly constant up to interval 10, increasing erratically thereafter. This tends to confirm what we found for the continuous model where the Weibull model with a scale parameter greater than one fitted better than the exponential. There are few observations for the later time intervals, which is why the hazard function is not smooth then.

## 7.7 Models when there is more than one episode

So far, our models have been restricted to the analysis of just one episode even though, as we discussed in Section 7.2, education careers generally consist of several episodes. To focus on one episode is not necessarily a drawback because sometimes it makes substantive sense to look separately at the first, second, and so on episodes. In our example, the first preschool episode is arguably more interesting than the subsequent episodes. On the other hand, it is a little inelegant to analyse episodes separately, and unsatisfactory to waste data by analysing only the first episode. There are in fact two pieces of information which we do not use if we restrict ourselves to the models described up to now. We ignore the fact that episodes can end in a number of different ways with transitions to different states, as well the possibility that there is more than one episode.

Let us first extend our models by taking account of the variability in the way

**Table 7.6** Discrete time models: estimates of discrete hazards

| Interval | Logit | Log-log |
|---|---|---|
| 1 | 0.14 | 0.14 |
| 2 | 0.19 | 0.19 |
| 3 | 0.09 | 0.09 |
| 4 | 0.10 | 0.10 |
| 5 | 0.19 | 0.18 |
| 6 | 0.16 | 0.16 |
| 7 | 0.11 | 0.11 |
| 8 | 0.19 | 0.19 |
| 9 | 0.25 | 0.25 |
| 10 | 0.19 | 0.19 |
| 11 | 0.26 | 0.26 |
| 12 | 0.28 | 0.28 |
| 13 | 0.29 | 0.29 |
| 14 | 0.54 | 0.53 |
| 15 | 0.14 | 0.14 |
| 16 | 0.17 | 0.17 |
| 17 | 0.40 | 0.40 |
| 18 | 0.67 | 0.67 |
| 19 | * | * |
| 20 | 0.99 | 0.99 |

*not enough data to estimate.

in which episodes can end; the situation we have previously described as competing risks. We could try to construct a model which simultaneously models the hazards for all possible transitions. This becomes rather difficult when there are a number of origins and destinations, as with the preschool data, because as the number of possible transitions becomes large, many will be observed rarely and so the estimation procedure will become imprecise. Instead, we can construct a series of models, one for each destination rather than for each transition. The original states are represented as a categorical explanatory variable as before, with the episode treated as censored if it ends with a transition to a state other than the chosen destination.

With our data, we consider two competing risks for the end of the first care episode: a change (in this case, a return) to care at home or a change to a different preschool environment (which could be a 'self transition' in the sense that the type of day care might not change; for example, a change of childminder). Out of the 234 first episodes, five were censored, 15 were not followed by a second episode of any description, which means that the children stayed in the same preschool environment until they started school, 72 children followed their first care episode by an episode at home and the remainder – 142 – followed the first preschool episode by another one.

Treating the 15 children with just one episode as making a transition to another preschool type (primary school), an admittedly arbitrary decision, then, when we fit two separate Cox PH models, we find a rather different pattern of results for the two competing risks. Table 7.7 gives the results and shows that the age at the start of the episode does not affect the hazard for those who change back to home care, but there is a marked start age effect – an increasing hazard – for those changing to a second preschool environment. The effects for mother's education are small for both risks but the effects of type of preschool (which we can think of as the original states) again vary by destination. The hazard is smaller for the nursery type than for all other types for transitions to preschool, whereas the hazard for the relative group is greater than the others for transitions to home care. Clearly, we could explore these competing risks in more detail by fitting parametric models for the hazard as we did in Section 7.5, and by fitting discrete time models. However, the analyses in this section demonstrate one way of dealing with transitions, and how, by extending the simple models of the previous sections, which did take account of origins but not of destinations, our conclusions can be modified.

**Table 7.7** Competing risks models: estimates from Cox models

| Explanatory variable | Transition to preschool | | Transition to home | |
|---|---|---|---|---|
| | Estimated regression coefficient | Standard error | Estimated regression coefficient | Standard error |
| Start age | 0.076 | 0.014 | 0.009 | 0.029 |
| Mother's education | 0.063 | 0.049 | −0.02 | 0.073 |
| Relative* | 0.72 | 0.18 | 0.93 | 0.32 |
| Childminder* | 0.60 | 0.15 | 0.45 | 0.31 |
| Nursery* | −0.45 | 0.23 | 0.01 | 0.38 |

*compared with playgroup as baseline category.

Finally, let us consider briefly the issue of multiple episodes. Essentially, we want to know whether episode durations within children are more alike than we would expect by chance. One way of getting at least an indication of whether this might be so is to regard the data as having two levels: level two is formed by the children and level one by the episodes within children. (And for datasets with children nested within schools then school could enter the analysis as level three.) We are now in the realm of multilevel models as described in Chapters 3 and 4. We could fit a simple two-level variance components model (see Section 3.3) to the log duration, i.e.:

$$\log t_{ij} = b_0 + u_j + e_{ij} \tag{7.14}$$

where $i$ represents episodes and $j$ children, with the implied assumption that the $e_{ij}$ are Normal or that duration has a log normal distribution. We would then be interested in whether the level two variance for the child mean durations ($\sigma_u^2$) was greater than zero. This approach might work well if there is little or no censoring but would need to be modified to take account of censoring. Censoring is not a major problem with the preschool data. There are a total of 1181 episodes distributed among 249 children, a mean of 4.74 episodes per child (dataset 7.3 on the disk). The level two, or between child, variation is 0.023 with a s.e. of 0.022 compared with a level one variance of 1.01 with a s.e. of 0.05. In other words, there appears to be little variation in mean duration between children. Therefore, we might be justified in treating all the episode durations as if they are independent, and hence in analysing all the episodes together. Another way of thinking about these particular data is to say that, as the observation period is essentially the same for all children, because it starts at birth and ends when the children start full-time school, then the number of episodes per child has a Poisson distribution with a mean of 4.74.

However, this is really only a small beginning to the analysis of multiple episodes. More sophisticated analyses would include explanatory variables, possibly with coefficients varying at level two, would account for censoring, would analyse hazards rather than durations and would allow for a range of distributional assumptions. At present, this is a relatively undeveloped area within event history analysis; some pointers can be found in Goldstein (1995, Chapter 9).

## 7.8 Conclusion

Event history analysis is an expanding field in applied statistics, fuelled by the increasing availability of longitudinal datasets. However, event history analysis has not been widely used in educational research up to now. In this chapter, I have covered the main methods without going extensively into the theory behind the methods. For readers wanting a good technical introduction to models for survival data, then Cox and Oakes (1984) is useful. The chapter by Tuma (1994) covers a range of technical and substantive issues. The book by Blossfeld et al. (1989) is written for social scientists, includes illustrations of how to use statistical packages for analysis, and shows how time dependent covariates can be incorporated into hazard models. The chapter in Aitkin et al. (1989) is detailed and oriented towards GLIM.

## Appendix 7.1

### Dictionary of terms used in event history analysis

Here we find brief descriptions, rather than rigorous definitions, of most of the terms we meet when we do or read event history analyses. Note, however, that many terms found in the statistical literature on the theory of stochastic processes are not included.

**Accelerated life model:** a model in which the observed *duration* (or *failure time*), or its log, is modelled directly, in contrast to *hazard* models.
**Alternating renewal process:** a *semi-Markov process* for two states.
**Bout:** *episode.*
**Censoring:** missing data in the sense of incomplete *episodes*. Censoring on the right arises because subjects are lost during a study, or because observation of the *event history* ends before the history itself. Censoring on the left arises when observation does not start at the beginning of the *event history* and so some events occur before observation does.
**Competing risks:** when there are two or more *events* which can determine the end of an *episode*. The occurrence of one of these *events* (e.g. death) means the other(s) cannot be observed (e.g. divorce).
**Continuous time event histories:** histories for which the exact timing of all *events* within the observation period is known.
**Cox model:** *proportional hazards model.*
**Cumulative hazard function:** minus the log of the *survivor function.*
**Discrete time event histories:** histories for which the order but not the exact time of all *events* in the observation period is known or histories for which *states* in the *state space* are known at certain fixed points in time (as in some panel studies).
**Duration:** the length of an *episode* in time units.
**Duration dependence:** when the *hazard* is a function of time so that an event becomes either less likely (negative duration dependence) or more likely (positive duration dependence) as time goes on.
**Episode:** the time between the start of an *event history* and an *event*, or the time between two successive *events*.
**Event:** the end of an *episode*, a change in *state*, a *transition*.
**Event history:** a sequence of *events*, in *continuous time* or in *discrete time*.
**Experience:** *state dependence.*
**Exponential distribution:** the distribution of *durations* (or *failure times*) when the *hazard* is constant, i.e. no *duration dependence*.
**Extreme value distribution:** the distribution of log *duration* if *duration* itself has a *Weibull distribution*.
**Failure time:** the time at which an *event* occurs (from industrial applications).
**Force of mortality:** *hazard* function.
**Frailty:** *unobserved heterogeneity.*
**Gompertz distribution:** the *hazard* is an exponential function of time.
**Gumbel distribution:** *extreme value distribution.*
**Hazard:** the rate at which *events* occur; the *risk* of an *event* occurring.
**Initial conditions:** differences between subjects at the start of observation, often caused by *left truncation*.

**Integrated hazard:** *cumulative hazard function.*
**Intensity matrix:** a matrix of *transition* rates, the rows of which sum to zero and the diagonal elements of which are the negative of the *hazard* (only applicable when there are multiple *events*).
**Lagged duration dependence:** when the *hazard* for the *episode* in question depends on the *duration* of the previous *episode* or episodes – only relevant for repeated *events*.
**Lagged occurrence dependence:** when the *hazard* for the *episode* in question depends on which *states* were occupied in the past – only relevant for multiple *events*.
**Left truncation:** when observation starts at different times for different individuals.
**Life table:** a representation of *survival probabilities* by age.
**Markov chain model:** a process in *discrete time* for which the *transition probability* from *state j* to *state k* does not depend on previous *states* occupied.
**Markov process:** a process in *continuous time* for which the *transition* from *state j* to *state k* does not depend on previous *states* occupied, i.e. a process which does not depend on history and for which *episodes* are *exponentially distributed*. It is a *Poisson process* generalized to more than one *state*.
**Non-parametric model:** a model which neither specifies a distribution for the *hazard* nor a functional form for the explanatory variables.
**Parametric model:** a model which specifies a distribution for the *hazard*.
**Partial likelihood:** a method of estimating parameters in a model, particularly a *proportional hazards model*, which is constructed from the order of observed *events*.
**Partially parametric model:** *semi-parametric model.*
**Poisson process:** a process for which the times between *events* are independently and *exponentially distributed* – only relevant for repeated single *events*.
**Proportional hazards model:** a model, sometimes known as a *Cox model*, for a *hazard* which varies with time but where the ratio of two *hazards* is independent of time, and therefore the distributional form of the *hazard* need not be specified. Hazards from a *Weibull distribution* are proportional.
**Renewal process:** a process for which the times between *events* are independently and identically distributed. A *Poisson process* is the simplest renewal process.
**Risk set:** the set of subjects at time $t$ yet to experience an *event* and whose *episodes* have not been *censored*.
**Sample path:** *continuous time event history.*
**Semi-Markov process:** a generalization of a *renewal process* to more than one state.
**Semi-parametric model:** a model, like a *Cox model*, which specifies a functional form for the explanatory variables but does not specify a distribution for the *hazard*.
**Spell:** *episode.*
**Starting conditions:** *initial conditions.*
**State:** a member of the *state space*.
**State dependence:** a general term covering *duration dependence*, *lagged duration dependence* and *occurrence dependence*.
**State space:** the set of distinct values, usually small, that an outcome variable can take.

**Stationarity:** *transitions* do not depend on time; a term usually applied to *Markov chains*.
**Stochastic process:** a general term to describe any process which evolves over time according to probabilistic laws.
**Survival analysis:** a term with origins in biostatistics describing methods for the analysis of *failure times*, often but not always applied to single unrepeated *events*.
**Survival curve:** *survivor function*.
**Survivor function:** calculated empirically from survival times to show how *failure* varies with age/time.
**Time homogeneity:** *stationarity*.
**Time varying covariates:** covariates (or explanatory variables) which change over time as the *event history* unfolds, as opposed to fixed covariates such as sex and ethnic group.
**Transition:** a change from one *state* to another.
**Transition probability:** the probability of a particular change in *state* between two points in time, which will depend on the history of the process if it is not *Markovian* and on time if it is not *stationary*.
**Transition rate:** a generalization of the *hazard* to multiple *events*.
**Unobserved heterogeneity:** unmeasured differences between individuals, including *initial conditions*, which can result in biased estimates of parameters. It can be modelled as a random effect.
**Waiting time:** *duration*.
**Weibull distribution:** a flexible distribution for the *duration* which can encompass positive and negative *duration dependence* (but not both), and includes the *exponential distribution* as a special case.

## Appendix 7.2

**GLIM code**

Note that the text in italics is comment and should not be typed.

*(A) GLIM code for Weibull analysis (Table 7.4)*
$slength 234$
$var id group epino start finish censor med diff2$
$fact type 4 (4)$
$data id group epino start finish censor type med diff2$
$dinp 'a:\ds71.dat'$
$calc censor = 1-censor$
$calc dur = finish-start$
$calc type = %if(type == 4,3,type)$
$calc type = %if(type == 5,4,type)$
*Here we input and call the Weibull macro.*
$input 9 80 weibull$
$mac model $endm$
$use weibull dur censor$
$mac model med$endm$
$use weibull dur censor$

$mac model med+type+start$endm$
$use weibull dur censor$
*Here we edit the Resplots macro in order to be able to save the standardized residuals.*
$edmac resplots$
10i $CAL RES=WK1_$
s
*Now we call the Resplots macro*
$use resplots censor$
$look RES$
*Now we fit an exponential model.*
$assign c=0$
$use weibull dur censor c$
$return$

*(B) GLIM code for discrete time analysis (Table 7.5)*
$slength 1384$
$fact type 4 (4)$
$var r nf rsi rss start med cons$
*These variables come from applying the SURV command in MLn; r is the response (0,1), nf is the number of failures, rsi is the risk set indicator, rss is the risk set size.*
$fact time 20$
*There are 20 discrete time intervals.*
$data r nf rsi time rss start med cons type$
$dinp 'a:\ds72.dat'$
$yvar r$error b cons$link g$
*This is the usual binomial model with a logit link, cons =1.*
$fit$
$eliminate time$
$fit$ :+type$ :+start$ :+med$
$disp e$
$elim$
$fit time$
$link c$
*This changes the model to have a log-log link. We then fit a new series of models as for the logit link.*

## Exercises

Answers to those exercises marked '#' can be found at the end of the book.

**1** The following data are times (in weeks) from arrival at secondary school until the pupil fails to hand in a piece of homework on time. Data points marked '*' are censored.
6 6 6 6* 7 9* 10 10* 12* 13 16 18* 18* 20* 22 23 25* 32* 32* 38* 45*
Obtain the survivor function and the cumulative hazard and comment on your findings.
(#)

**2** Using dataset 7.1 on the disk, extend the analysis of log durations (i.e. Table 7.2) to include the group variable (returners and non-returners), and interactions between start age and preschool type.

**3** Obtain the cumulative hazard function for the third episodes ($n=221$) in dataset 7.3 on the disk. Note that these are the second episodes after the period at home *after* birth.

**4** Fit a Cox proportional hazards model to the third episodes in dataset 7.3, on the disk with age at start and preschool type as explanatory variables. Note that there are now seven preschool types; you should recode the small number of type=3 to be type=2. These types are therefore home, relative, childminder, nursery, playgroup, and nursery school/infant school. Compare the results with those from the same model fitted to the first episodes (Table 7.3).
(#)

**5** Obtain an appropriate residual plot for the logistic regression of the discrete time data (dataset 7.2 on the disk) given in Table 7.5. Comment on your results.

# 8
# Modelling Educational Data: Further Issues

## 8.1 Introduction

The previous six chapters have covered, in a practical and example-driven way, the most frequently used statistical models in educational research. This final chapter is a little different in that it covers more briefly a number of more advanced topics. More detailed treatments of these topics can be found in the references provided. But here we see something of the contexts in which these models can be useful and some examples of where, rather than precisely how, they have been applied. The chapter is grouped into three main sections. In the first section, some of the modelling issues arising from problems caused by *measurement error*, especially measurement error in explanatory variables, are discussed. This is followed by a section covering two further topics in multilevel modelling which are of particular interest to educational researchers – *cross-classified models* and *multivariate models*. Finally, we meet some of the problems in the analysis of categorical data which have received a lot of attention from statisticians in recent years. These include modelling *multilevel categorical* data.

## 8.2 Measurement errors

### 8.2.1 Correcting regression models

If we think back to the variables in our examples in Chapters 2, 3 and 4 – the chapters concerned with models for continuous responses – many of these variables would have been measured with error. For example, we know that attainment measures are not wholly reliable, with pupils' performance on tests varying from item to item, from day to day, and from tester to tester. The measure of curriculum coverage is also likely to be an imperfect indicator of the underlying variable as it was constructed from teachers' responses to a set of items. We can formalize this notion of imperfect, or unreliable, measurement by constructing a statistical model for measurement error. Many models have been proposed, some relatively simple and some complex. The most simple, and the most widely used model, separates any observed score on a continuous variable into two components:

$$S_i = T_i + E_i \qquad (8.1)$$

where $S_i$ is the observed score on the variable of interest for individual $i$, $T_i$ is a notional *true score* for individual $i$ which is, by definition, unobservable, and $E_i$ is measurement error. Observed and true scores are sometimes called scores on *manifest* and *latent* variables respectively. Textbooks on psychometrics, such as Lord and Novick (1968), are the best source for detailed discussions about measurement error models.

The important parts of this model are the assumptions we make about $E_i$. First, we assume that its mean is zero. In other words, we assume that our measure, although fallible, is not systematically inaccurate, or biased. If the measure were biased, and if we knew the size of the bias, then we could just subtract the bias from the observed scores, leaving our model essentially unchanged. However, it is rare to know the size of a bias and there is nothing we can do, when we are analysing data, to correct for unknown measurement biases. Instead, we must rely on the best efforts of those who develop measures to keep biases to a minimum. We also assume that the size of the measurement error, $E_i$, is independent of the true score $T_i$, and that its variance, $\sigma_E^2$, is constant. These assumptions are illustrated by Figure 8.1. It is also sometimes convenient to assume that the measurement errors are Normally distributed.

It is very important to remember that this decomposition of observed scores into true scores and errors is a statistical model often referred to as a *measurement model*, with rather strong assumptions which are not easy to test. Also, all the potential sources of measurement error – from day to day, from tester to tester and so on – are lumped together in $E_i$, rather than being considered separately, as they are in *generalizability theory* (see, for example, de Gruijter and van der Kamp, 1991). Nevertheless, it is a parsimonious model which can be

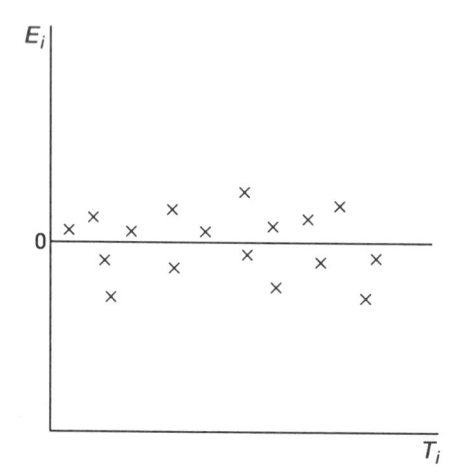

**Figure 8.1** Scatterplot of measurement error and true score

## 8.2 Measurement errors

incorporated into regression models in a relatively straightforward way. However, the measurement model is only of any practical use when we have some extra information. Either we need to know, or to have an estimate of, the error variance $\sigma_E^2$, or we need to know the ratio of the true score variance to the observed score variance. This latter ratio is usually called the *reliability* of the measure, with a reliability of one corresponding to $\sigma_E^2=0$.

The best known effect of measurement error is the so-called *attenuation effect* on correlation coefficients. If $x$ and $y$ are both imperfectly measured then the correlation between their true values, $X$ and $Y$, is equal to the observed correlation divided by the square root of the product of the two reliabilities, $r_{xx}$ and $r_{yy}$:

$$\rho_{XY}=\rho_{xy}/\sqrt{(r_{xx}r_{yy})} \qquad (8.2)$$

so if, for example, the observed correlation is 0.5, $r_{xx}$ is 0.8 and $r_{yy}$ is 0.6, then the true correlation, 0.72, is quite a lot higher. This is a situation in which we must take account of measurement error in both $x$ and $y$. However, if we focus on the simple regression model of Chapter 2:

$$y_i=a+bx_i+e_i \qquad (8.3)$$

then it is the measurement error in $x$ which is of most importance. The major effect of measurement error in the response is to set an upper limit to the amount of variance that can be explained by any combination of perfectly measured explanatory variables. If the reliability of $y$ is 0.8 then only 80% of its variance can be explained. However, if $x$ is measured imperfectly then the least squares estimate of $b$ (p. 12) is biased. Rather than $\hat{b}_{\text{OLS}}$, the unbiased estimate of the effect of $x$ on $y$ is $\hat{b}_{\text{OLS}}/r_{xx}$. In the same way, the unbiased estimate of the constant term is $\bar{y}-(\hat{b}_{\text{OLS}}/r_{xx})\bar{x}$.

Providing we have a good estimate of the error variance, or the reliability, of a single explanatory variable, then correcting a simple regression is straightforward. For example, the relation between LOGMATH and ZCURRIC in Chapter 2 was, without correction for measurement error in $x$:

$$\text{LOGMATH}=3.29+0.23\text{ ZCURRIC} \qquad (8.4)$$
$$(0.047)\ (0.048)$$

Assuming a reliability of 0.9 for ZCURRIC, then, after correction, the regression becomes:

$$\text{LOGMATH}=3.29+0.26\text{ ZCURRIC} \qquad (8.5)$$
$$(0.048)\ (0.037)$$

Note that the estimate for ZCURRIC increases, which is bound to happen, and the intercept has not changed, because the mean of $x$ is zero. Note also that, in this case but not always, the standard error of the regression coefficient is smaller after correction.

Correcting a multiple regression is more complicated because the simple results of the previous paragraph are no longer available. Consider the model whose estimates are given on p. 21

$$\text{LOGMATH}=3.29+0.11\text{ ZCURRIC}+0.20\text{ ZMATH1} \qquad (8.6)$$
$$(0.04)\ (0.05)\qquad\qquad(0.05)$$

Let us see what happens when we correct for measurement error in the two explanatory variables. We consider four combinations of reliability, with values of 0.7 or 0.9 attached to each of the explanatory variables. Table 8.1 gives the results. They were obtained from the program EVCARP (Schnell et al., 1988), and are based on theoretical work described in, for example, Fuller (1991).

Table 8.1 shows just how much our conclusions can be affected by the amount of measurement error. This is particularly so for the effect of curriculum coverage on mathematics progress; the effect is essentially zero if the reliability of the mathematics test is 0.7, regardless of the reliability of the curriculum coverage measure itself. Note also that the standard errors vary considerably according to the assumptions we adopt about reliabilities.

**Table 8.1** Effects of correcting for measurement error on regression estimates and standard errors

| Term | Reliabilities | | | | |
|---|---|---|---|---|---|
| | (1,1) | (0.9, 0.9) | (0.9, 0.7) | (0.7, 0.9) | (0.7, 0.7) |
| Intercept | 3.3 (0.04) | 3.3 (0.04) | 3.3 (0.05) | 3.3 (0.04) | 3.3 (0.04) |
| ZCURRIC | 0.11 (0.05) | 0.11 (0.06) | 0.013 (0.11) | 0.17 (0.10) | 0.026 (0.22) |
| ZMATH1 | 0.20 (0.05) | 0.23 (0.07) | 0.37 (0.11) | 0.18 (0.10) | 0.36 (0.18) |

Table 8.1 shows how important it is to get accurate estimates of the reliabilities of explanatory variables. A full analysis of the effects of measurement error on regressions needs to take account of the precision of the reliability estimates, and also needs to consider the effects of error covariances. For example, it is possible for measurement errors of the explanatory variables to be correlated across individuals if the same interviewer or tester were used for more than one measure.

### 8.2.2 Latent variable models

The previous sub-section showed how important it is to correct for measurement error in those situations for which we have an external estimate of the reliability or error variance of a manifest explanatory variable. However, there is another approach which we can adopt when we have more than one measure, or indicator, of a latent variable. This approach is often referred to as a structural equations approach to measurement error and has its roots in the statistical technique of factor analysis. Quite complicated structural equation models can be built up, including several linked, or causally related, latent variables, as well as many indicators. Nowadays, there are several computer packages to choose from to do this kind of analysis, the best known being LISREL, EQS and AMOS. Hox (1995) gives a comparative review of these three packages.

### 8.2.3 Misclassification and latent class models

Up to now, we have been concerned with measurement error in variables with interval or continuous scales. However, the binary and categorical responses which are the focus of Chapters 5, 6 and 7 will also, in many cases, be measured with error. For example, it is probable that all the variables in Table 5.4 (college plans, socio-economic status, IQ and parental attitudes) were measured imperfectly, with the likelihood of variation in the way questionnaires were administered, together with the difficulty of measuring intentions and attitudes with just a single variable or indicator. In Table 5.5, we related mothers' views about their children's performance at two points in time. The off-diagonal cells represent cases of observed change but these could just be cases of misclassification. In other words, the way the question was asked at the two occasions, the mood of the mothers at the time of interview and, perhaps, most importantly, the difficulty of measuring the underlying variable – mothers' satisfaction with their children's school progress – with just a single question, could all lead to the appearance of change where none really existed. Moreover, it is also possible that the diagonal cells, representing no change, include cases where there had been a genuine change that was not picked up.

The measurement error model of Section 8.2.1 for continuous variables cannot be applied when the data are categorical. For example, for any truly binary variable, the measurement error must be positive when the value is 0 and negative when the value is 1 so $T_i$ and $E_i$ cannot be independent. Statisticians have therefore developed models known as latent class models to try to deal with this kind of measurement error. These models are often statistically and computationally complex but the basic idea behind them, first formulated by the sociologist Lazarsfeld in the 1950s, can be stated quite simply. Suppose we have a set of binary indicators measured on a sample of individuals. Suppose also that we believe that the individuals can be placed into a number of unobserved, or latent, classes, usually no more than three or four. Our observed data are a series of 'yes' and 'no', or 0 and 1, responses to each item for each individual. However, we recognize that these responses are an outcome of a process in which the probability of an individual saying 'yes' to an item will vary according to the latent class that the individual belongs to which. A 'yes' response to an item is much more likely if the individual is in class A, say, rather than in class B, C etc, but the probability of a 'yes' is not necessarily zero if the individual is not in class A. Latent class modelling has three main aims in this situation: to establish how many latent classes there are, to estimate the proportions of the population in each class, and to estimate the probabilities of a 'yes' response for each item within each class. In addition, we sometimes want to estimate the probability of being in a particular class for each individual.

An application of latent class modelling to educational data is given by Aitkin et al. (1981), in their well known re-analysis of the Bennett data (see p. 34). A set of 38 binary items about teaching methods was presented to 468 teachers. On the basis of the observed responses, both two- and three-class models were derived, and these latent classes were interpreted in terms of teaching styles. A sub-sample of 37 teachers was analysed in more detail and 26 of these were shown to belong to one of the three latent classes with a probability of at least 0.9. For some items, the probability of a 'yes' reply did not vary much across the latent classes, for others it did. For example, the probability of a 'yes' reply to the item 'pupils not allowed freedom of movement in classroom' was estimated to be 1 for latent class one but only 0.53 for latent class two.

Returning to Table 5.5, we do not have enough information there to separate observed change from true change. However, with more indicators and more occasions of measurement, it would be possible to estimate a latent class model which provided estimates of the true turnover or transitions over time. An introduction to latent class modelling is given by Hagenaars (1990, Chapter 3).

### 8.2.4 Measurement errors in event histories

In event history analysis, there are two sources of error: in the measurement of the process itself and in the measurement of the explanatory variables. Errors inherent in retrospective data collection – which is almost unavoidable when collecting event histories – dominate the first source. First, episodes, especially short episodes, can be forgotten altogether. Secondly, transitions may not be accurately dated, with a tendency to 'telescope' events from the distant past into the more recent past. Thirdly, dates may be observed to be 'heaped' at particular times, such as seven days, 12 months and so on. Some evidence for heaping comes from the preschool data, analysed in Chapter 7. Table 8.2 shows more episodes ending at 18, 24, 36 and 48 months than at one month either side. One possible solution to the problem of heaping is to smooth the data around the point at which they are heaped. Fourthly, histories may not be 'seamless' in the sense that the current state reported at an interview may not be the same as the equivalent state recalled at a subsequent interview. Finally, as we have seen with the preschool data, there can be ambiguity about how a transition is defined so that it is not always clear whether or not a change of state has occurred. All of these measurement issues can affect the distributions of durations and hazards but there are few discussions of them in the literature.

**Table 8.2** Evidence for 'heaping': number of episodes ending within one month of chosen dates

| Episode end (months) | Deviation −1 | 0 | +1 |
|---|---|---|---|
| 18 | 8 | 17 | 3 |
| 24 | 10 | 15 | 6 |
| 36 | 7 | 28 | 24 |
| 48 | 13 | 16 | 7 |

If we adopt an accelerated life model for log duration, then we can correct for measurement errors in continuously measured explanatory variables using the methods described in Section 8.2.1. However, little is known about the effects of these kinds of measurement errors on hazard models.

## 8.3 Extensions of multilevel modelling

Some basic multilevel models for educational data were described in Chapters 3 and 4. However, it is not unusual to meet educational data which do not fit easily into those basic models. Here, I briefly describe two extensions: to cross-classified structures and to multivariate data. Another extension – multilevel modelling of categorical data – is covered in Section 8.4.

### 8.3.1 Cross-classified multilevel models

Consider the following research question: can we explain variation in pupils' public examination results at age 16 in terms both of the secondary school and the primary school they attended? If pupils from a single primary school all go

## 8.3 Extensions of multilevel modelling 143

to the same secondary school, then we still have a hierarchical structure with three levels as shown in Figure 8.2. We can estimate this model using the methods described in Chapter 3, ending up with estimates of variation in examination results between primary schools within secondary schools (i.e. level 2) and between secondary schools (level 3). However, it is much more common for primary school leavers to be dispersed across several secondary schools, as indicated in Figure 8.3. No longer do we have a hierarchical structure, instead we have a cross-classification at level two, with pupils still defining level one. In these situations, we estimate two random effects at level two, possibly more if

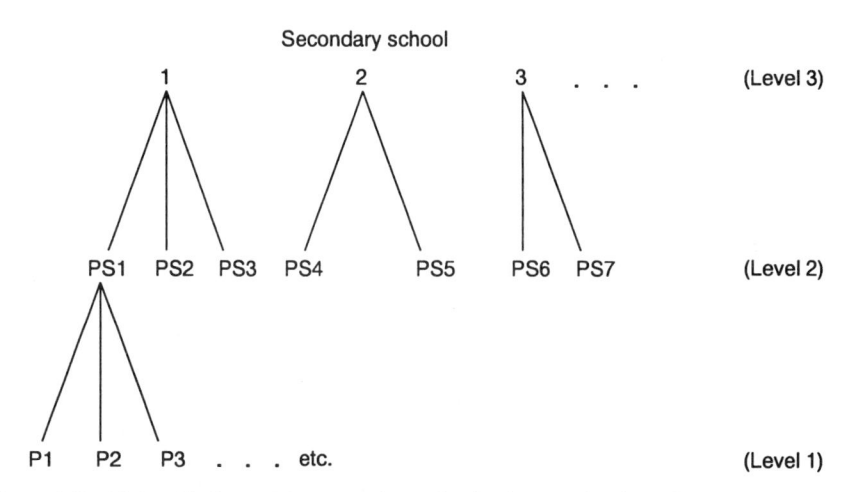

**Figure 8.2** Hierarchical structure: secondary school, primary school, pupil

| Primary School | Secondary school |   |   |     |
|---|---|---|---|---|
|   | 1 | 2 | 3 | ... |
| 1 | 40 | 5 | 5 |   |
| 2 | 10 | 20 | 0 |   |
| 3 | 1 | 30 | 5 |   |
| 4 | 5 | 5 | 40 |   |
| 5 | 50 | 0 | 0 |   |
| 6 | 0 | 30 | 0 |   |
| 7 | 0 | 0 | 25 |   |
| ... |   |   |   |   |

**Figure 8.3** Cross-classified structure: secondary school by primary school, pupils as cell counts

there are explanatory variables in the model. In other words, we estimate a between secondary school and a between primary school variance at level two. The limited evidence that has been acquired on the above research question suggests that between primary school variation in examination results can be more important than between secondary school variation (Goldstein and Sammons, 1997).

The computational aspects of estimating cross-classified multilevel models are considerable. We see in Figure 8.3 that pupils are not uniformly dispersed across the three secondary schools; secondary school one takes most of its pupils from primary schools one and five, secondary school three from primary school four and so on. This is not an unusual situation, and we can exploit this imbalance in the analysis to reduce the computational burden. More details of how to analyse different kinds of cross-classifications can be found in Goldstein (1995, Chapter 8).

### 8.3.2 Multivariate multilevel models

If we think back to Chapter 3, we partitioned the variability in an overall measure of curriculum coverage into variability between and within classrooms. However, the overall measure is made up of measures for three sub-scales, based on the three Attainment Targets (ATs) at Key Stage One. These ATs are defined in Plewis and Veltman (1996). Consequently, it is of some interest to find out whether the correlations between these three sub-scales are the same *between* classrooms as they are *within* classrooms. This is an example of a simple multivariate multilevel model which generates three levels: classrooms (level 3), pupils (level 2) and AT sub-scales (level 1). There are clear analogies between a multivariate multilevel model and a multilevel model for repeated measures, where the occasions define level one. However, with a multivariate model, we do not do any modelling of the level one variance.

Table 8.3 gives the estimated correlations for the two levels of interest. We see that the within teacher correlations are considerably higher than the between teacher correlations. In other words, pupils in a class who cover a lot or a little of one AT also cover a lot or a little of the other ATs. However, teachers who cover a lot or a little of one AT do not necessarily cover a lot or a little of the other ATs. Looking at correlations both between and within higher level units can give us more insight into our data. We could go on to elaborate this model by including explanatory variables and seeing whether the relationships with the explanatory variables vary from sub-scale to sub-scale, and from school to school. Duncan et al. (1995) give an example of this kind of multivariate analysis.

**Table 8.3** Correlations within and between classrooms

|     | Within classrooms | | Between classrooms | |
| --- | --- | --- | --- | --- |
|     | AT1 | AT2 | AT1 | AT2 |
| AT2 | 0.90 |      | 0.61 |      |
| AT3 | 0.79 | 0.80 | 0.67 | 0.47 |

## 8.4 Modelling structured categorical data

We have seen, in Chapters 2 and 3 and again in the two previous sections of this chapter, how to model hierarchical structures which are so common with educational data. However, all these models have only been suitable for continuous responses. Many of the examples we used in Chapters 5 and 6 – the chapters dealing with binary and categorical data – were generated from studies with a hierarchical structure which was not made explicit there. For example, Table 5.4 was constructed from responses to a questionnaire given to pupils in an unknown number of schools; Table 6.5 was based on just three schools but was extracted from a much larger sample with many schools. Multilevel modelling of categorical data is both less advanced in its development and use, and theoretically more difficult than multilevel modelling of continuous responses, and therefore only a brief introduction is appropriate here.

When we ignore the hierarchical structure linked to categorical responses, we create a problem which is often referred to in the technical literature as *overdispersion* or *extra variation*. In other words, our usual distributional assumptions – Binomial for binary responses and Poisson for categorical responses – no longer hold. Within each cell of a contingency table like Table 5.4, there is extra variation beyond that generated by the Binomial or Poisson distributions, arising from the fact that there is additional between school variation. Ignoring this extra variation has two effects – the standard errors attached to the estimates from a model are too low, and the goodness of fit statistics (the deviances) are misleading. In particular, the deviance can often indicate that a main effects model does not fit well when, in fact, this lack of fit arises not from a need to include higher-order interactions but from the extra variation, generated by the hierarchical structure.

There are at least two classes of methods which have been introduced in recent years to deal with this problem. The first goes under the name of *population averaged* models. With these models, the focus is on the fixed rather than on the random effects, and these fixed effects are estimated on the basis of a more general covariance structure for the response than the one defined by the usual distributional assumptions. *Generalized estimating equations* (GEEs) are often used to estimate these models – see Diggle *et al.* (1994).

The second class are often known as *subject specific* or *unit specific* models. Essentially, these are multilevel models for categorical data, and they are especially useful when we want to measure and to explain the variation in any random effects. Suppose we have a binary response, a single explanatory variable and suppose also that the data come from a population with two levels – pupils and schools say. Then our two level logistic regression model is:

$$\text{logit } P_{ij} = b_0 + b_1 x_{ij} + u_j \qquad (8.7)$$

where $i$ indexes pupils and $j$ indexes schools, as before. The $u_j$ term, which was not of course in the model in Chapter 5, represents the level two effects. These $u_j$ are assumed to be Normal with constant variance. As we did in Chapter 5, we can write the model in terms of the observed probabilities, rather than the expected probabilities. It is then:

$$P_{ij} = \exp(b_0 + b_1 x_{ij} + u_j)/(1 + \exp(b_0 + b_1 x_{ij} + u_j)) + f_{ij} \qquad (8.8)$$

so that the level two term is in the non-linear part of the model and the level one term, $f_{ij}$, is not. The usual assumption for the $f_{ij}$ is that they are Binomially distributed. The model is reasonably easy to specify, and interpretation of the results is basically the same as for linear models for continuous responses. The problems come with the way the model should be estimated, and the properties of the estimation procedure. The different approaches are described in Goldstein (1995, Chapter 7) and in Diggle *et al.* (1994). There are now a number of packages which will estimate multilevel generalized linear models such as MLn (Rasbash and Woodhouse, 1995), which has specialist macros, HLM (Bryk *et al.*, 1996), EGRET (1993), and BUGS (Spiegelhalter *et al.*, 1995). Paterson (1995) shows how to specify and interpret a two-level logistic regression using MLn.

## 8.5  Concluding remarks

This chapter has maintained the modelling theme of the earlier chapters by introducing some more advanced topics. The way in which measurement error can affect model estimates is especially important, but is rarely considered in quantitative analyses of educational data. This is partly because we do not always have good data on the extent of measurement error, partly because the methods can be complex, particularly for categorical data, and partly because the methods are not built into the popular statistical packages. Moreover, we are only just beginning to understand how measurement error can affect the estimates from multilevel models (Woodhouse *et al.*, 1996). We have also seen, or at least glimpsed, how multilevel models can be extended from the relatively straightforward models for hierarchically structured continuous responses to the more complicated situations of cross-classified structures, multivariate structures and models for categorical data.

To conclude, let us return to the arguments in favour of statistical modelling put forward in the first chapter. In nearly all our examples, we have related a response to one or more explanatory variables and, by so doing, we have seen how our models can give us not only a better description, but also a better understanding of the underlying educational processes. We have seen how important assumptions are when we build models, and how these assumptions need to be checked by carrying out different kinds of diagnostic tests. In each of Chapters 2 to 8, we have met different ways of modelling educational data. However, each model has been linked to previous models in order to bring out the essential unity of statistics. Finally, although the mathematics behind the models has generally been kept in the background, we have seen how, by writing down the models as equations, we can expand and develop our ideas about how the underlying processes can be modelled.

Critics of statistical modelling sometimes assert that using models is mystifying, taking researchers away from their data, and making it impossible for users of research to understand what the researchers have done. This book has shown that there is no need for statistical modelling to be a mysterious process. We have seen that analysing data by using statistical models is a public activity which can be reproduced and extended by others. It also works best when it is collaborative; researchers and statisticians working together to find a joint solution to an educational problem. We avoid mystery by carefully explaining what we have done and what assumptions we have made, and by committing ourselves to making the results of research accessible to all its potential users.

This book will have achieved one of its aims if it leads researchers to say 'this is the question I want to address: given the data available to me, what is the most appropriate statistical model I should use to answer it?' It will have achieved another aim if you then turn to one of the chapters to find out just how to use it. And it will have achieved its final aim if you are then able to interpret the fitted model in the light of your original research question.

# Answers to Selected Exercises

## Chapter 2

**2** (a) LOGMATH $= 3.43 - 0.26\, d_1 - 0.18\, d_2$

where $d_1$ is 1 for school two and 0 otherwise and $d_2$ is 1 for school three and 0 otherwise. The standard errors are 0.10, 0.16 and 0.13 respectively.

(b) LOGMATH $= 3.43\, d_1 + 3.17\, d_2 + 3.25\, d_3$

where $d_1$ is 1 for school one and 0 otherwise, $d_2$ is 1 for school two and 0 otherwise, and $d_3$ is 1 for school three and 0 otherwise. The standard errors are 0.10, 0.12 and 0.09 respectively.

In (b), we immediately see that the estimated school means are significantly different from zero but this representation is much less informative about whether the estimated mean differences between the schools are statistically significantly different from zero. The $R^2$ and $F$ statistics from model (b) – technically known as *regression through the origin* – are misleading and therefore it is usually better not to use this way of representing categorical explanatory variables.

**3** Suppose the response is a measure of self-esteem of pupils in reception classes and the two explanatory variables are sex (B,G) and nursery experience (Yes, No). Suppose the means can be represented diagrammatically as follows:

## 150  Answers to exercises

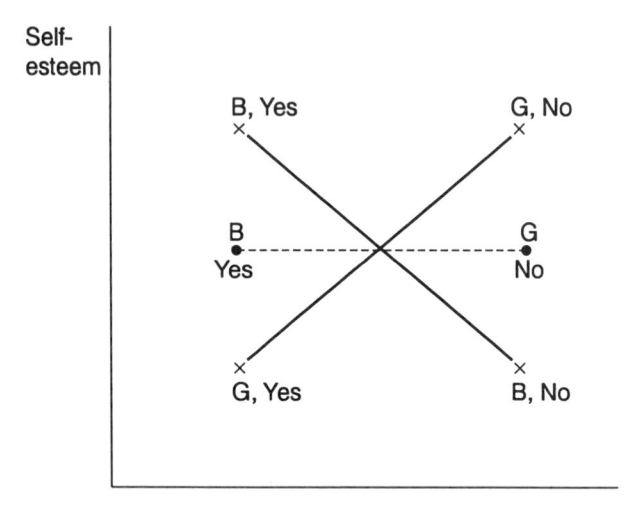

[Figure: Self-esteem plot with B,Yes (top-left) and G,No (top-right) marked with ×; B Yes and G No marked with • on the middle dotted line; G,Yes (bottom-left) and B,No (bottom-right) marked with ×.]

In other words, asuming the four groups are of equal size, there are no overall sex differences in self-esteem, nor does self-esteem vary by nursery experience (the dotted line). However, in this fictitious example, there is an advantage to nursery experience for boys and an equal disadvantage for girls.

## Chapter 3

**1**

Attainment by progress residuals: school

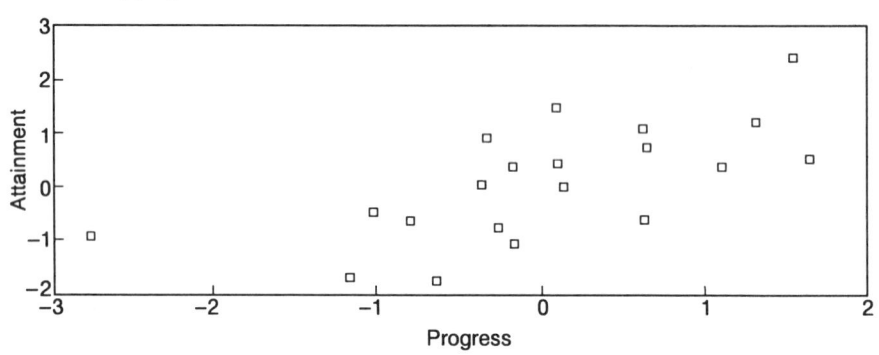

[Scatter plot of Attainment (y-axis, −2 to 3) vs Progress (x-axis, −3 to 2).]

Pearson correlation = 0.64 (note outlier for progress)
Spearman rank order correlation = 0.66

Attainment by progress residuals: teacher

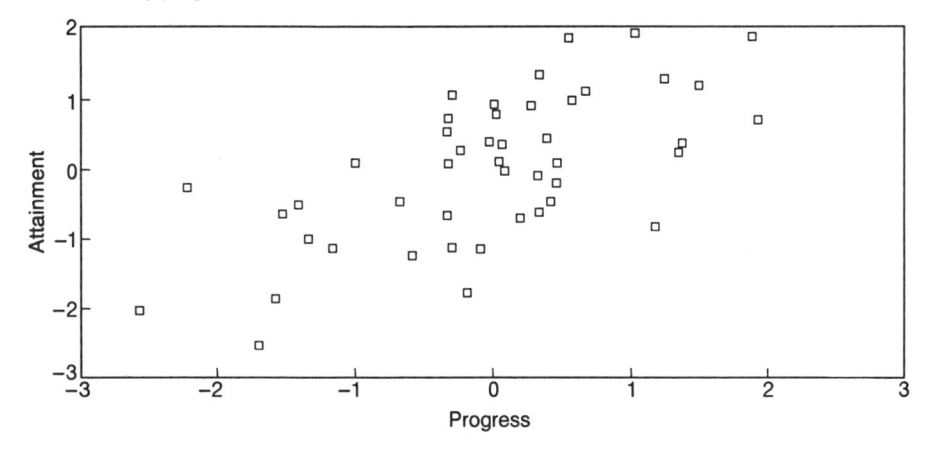

[scatter plot: Attainment vs Progress]

Pearson correlation = 0.65
Spearman correlation = 0.61

We see that although schools and classrooms with higher levels of reading attainment tend to have higher levels of reading progress, the correlation is far from perfect. The correlations are essentially the same for schools and classrooms. Hence, there are strong grounds for arguing that institutional rankings based solely on outcomes do not properly represent institutions' achievements.

3  (a) $ZCURRIC_{ij} = b_{0j} + e_{ij}$

   $\hat{\sigma}^2_{u(0)} = 0.32\ (0.10)$

   $\hat{\sigma}^2_e = 0.66\ (0.048)$

   $\hat{\rho} = 0.33$

   Deviance = 1041

(b) $ZCURRIC_{ij} = b_{0j} + b_1 ZMATH1_{ij} + e_{ij}$

   Deviance = 903; improvement = 138 for 1 df

   $\hat{\sigma}^2_{u(0)} = 0.23\ (0.075)$

   $\hat{\sigma}^2_e = 0.47\ (0.034)$

   $\hat{\rho} = 0.33$ – unchanged from (a) so ZMATH1 accounts for the same proportion of between and within classroom variance.

(c) $ZCURRIC_{ij} = b_{0j} + b_{1j} ZMATH1_{ij} + e_{ij}$

   Deviance = 880; improvement = 23 for 2 df ($p < 0.001$)

   $\hat{\sigma}^2_{u(0)} = 0.19\ (0.064)$

   $\hat{\sigma}^2_{u(1)} = 0.060\ (0.028)$

   $\hat{\sigma}^2_{u(0)u(1)} = -0.046\ (0.031)$

   $\hat{\sigma}^2_e = 0.43\ (0.032)$

Hence, between classroom variance in curriculum coverage varies according to the value of initial mathematics attainment and therefore a single value of ρ is not helpful. When ZMATH1=0 (i.e. the mean), $\hat{\rho}=0.31$.

Between classroom variance = $0.19 - 0.092$ ZMATH1 $+ 0.06$ ZMATH1$^2$.

## Chapter 4

**1** On the random effects side, we see that the linear growth rates vary from school to school, the 95% coverage interval being 0.88 to 1.12. The correlation between level and growth rate at the school level is high (0.95). The level three variance increases with age as:

$0.074 + 0.032$ AGE $+ 0.004$ AGE$^2$.

The 95% coverage interval for the linear growth rate for pupils is 0.73 to 1.27. There appears to be little between pupil variation in quadratic growth. The correlation between level and linear growth is 0.94, between level and quadratic growth it is –0.38 and between linear and quadratic growth it is –0.02.

There is heteroscedasticity at level one. The level one variance increases with age as:

$0.18 + 0.025$ AGE $+ 0.008$ AGE$^2$.

Turning to the fixed effects, we see that the two main effects and the interaction are all greater than twice their standard errors. This is also true for the interactions between these three terms and both AGE and AGE$^2$. White boys move ahead of the other three groups during infant school by between 1/3 and 1/2 SD units. The differences in relative growth during junior school are smaller, although black boys fall behind white boys by 1/10 SD units.

## Chapter 5

**1** (a)

|  |  | Child aged 14 years |  |  |
|---|---|---|---|---|
|  |  | Problem | No problem | Total |
| Child aged 12 years | Problem | 8 | 17 | 25 |
|  | No problem | 2 | 23 | 25 |
|  | Total | 10 | 40 | 50 |

Fall in problems of 30% (from 50% to 20%).

$M^2 = 225/19 = 11.84$; distributed as chi-square with 1 df; $p<0.01$.

(b)

|  | London | New York | Total |
|---|---|---|---|
| Problem | 10 [0.2] | 15 [0.25] | 25 |
| No problem | 40 [0.8] | 45 [0.75] | 85 |
| Total | 50 | 60 | 110 |

Slightly more problems in New York (from conditional probabilities).
$\hat{\alpha}=0.75$ (s.e. $=0.35$); $\hat{Q}=-0.14$ (s.e. $=0.23$) so little association between city and homework problems.

Expected values under null hypothesis of independence:

11.4  13.6
38.6  46.4

$X^2=0.40<3.84, p>0.5$.

**5** Omitting the outlier for 'time reading aloud', and so, for $n=192$, and coding sex as 1 for boys and 0 for girls:

logit $P=-2.24+0.0073$ READING ALOUD $-0.23$ SEX
    (0.36) (0.0031)                    (0.46)

so no evidence for a sex difference in the probability of being a good reader.

Introducing the interaction between sex and reading aloud:

logit $P=-2.10+0.011$ READING ALOUD $-0.60$ SEX $-0.0064$ SEX.RALOUD
    (0.38) (0.0051)                     (0.59)      (0.0064)

so no evidence for the interaction either.

## Chapter 6

**1** There were four degrees of freedom for the usual test of independence for Table 6.1. Creating three structural zeros leaves just one df. The test of the hypothesis of quasi-independence gives a $\chi^2$ value of 1.93 ($p>0.16$), and so the evidence is consistent with the view that teachers' expectations ratings are independent *if* they are not in agreement.

**3** Logged continuation odds for Table 6.6: 0.86; −0.08 (boys) and 1.59; 0.19 (girls).

Cumulative logits for Table 6.7: −0.64; 2.16 (boys) and −2.26; 1.59 (girls).

## Chapter 7

**1**

| $j$ | 1 | 2 | 3 | 4 | 5 | 6 | 7 | |
|---|---|---|---|---|---|---|---|---|
| $t_j$ | 6 | 7 | 10 | 13 | 16 | 22 | 23 | Change points / Survival times |
| $r_j$ | 21 | 17 | 15 | 12 | 11 | 7 | 6 | Risk set |
| $d_j$ | 3 | 1 | 1 | 1 | 1 | 1 | 1 | Non-survivors |
| $h_j$ | 0.14 | 0.06 | 0.07 | 0.08 | 0.09 | 0.14 | 0.17 | Hazard |
| $1-h_j$ | 0.86 | 0.94 | 0.93 | 0.92 | 0.91 | 0.86 | 0.83 | |
| $S_j$ | 0.86 | 0.81 | 0.75 | 0.69 | 0.63 | 0.54 | 0.45 | Survivor function |
| $H_j$ | 0.15 | 0.21 | 0.29 | 0.37 | 0.46 | 0.62 | 0.80 | Cumulative hazard |

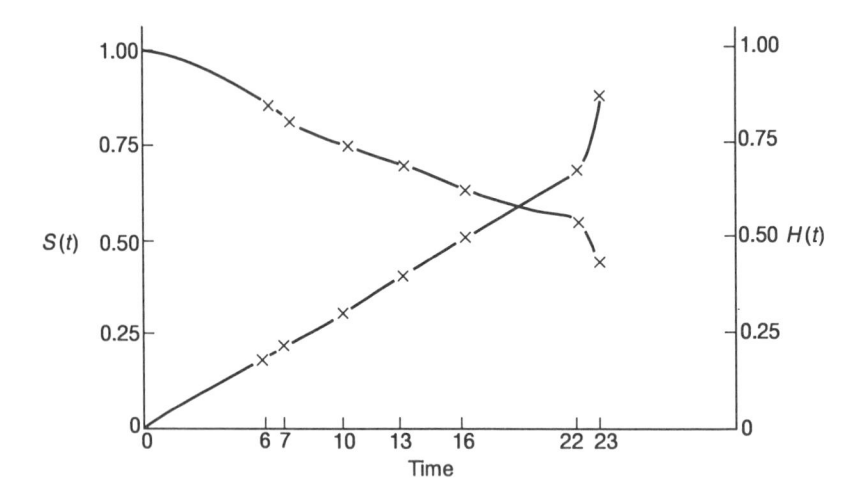

So we see that the sample cumulative hazard (the right-hand side scale) deviates from a straight line as time increases, suggesting that a constant (exponential) hazard may not be appropriate. We see from the data that the actual hazard declines and then increases, suggesting a 'bath tub' shape. However, the amount of data is small.

4   Eight of the 221 episodes are censored. The estimates are:

| Explanatory variable | Estimated regression coefficient | Standard error | 95% confidence limits for exp(b) |
|---|---|---|---|
| Start age | 0.045 | 0.0072 | 1.03; 1.06 |
| Home | 0.82 | 0.30 | 1.25; 4.12 |
| Relative* | 0.68 | 0.32 | 1.05; 3.68 |
| Childminder* | 0.62 | 0.28 | 1.09; 3.20 |
| Nursery* | −0.35 | 0.40 | 0.32; 1.55 |
| Playgroup* | 0.13 | 0.34 | 0.58; 2.24 |

*compared with nursery school/infant school as the base category.

The effect for 'start age' is similar to that found for the second episodes in Table 7.3. The effects for 'type' are not directly comparable because there are more types to consider for the third episodes, including 'home'. It does appear that the hazards are higher for 'home' care (the first three categories) than they are for institutional experiences (the final three categories), and this is consistent with Table 7.3.

Therefore, we might combine the 'type' groups to make a 'home' group and an 'institution' group, and we also consider the possibility of an interaction between type and start. The fitted model is then:

$$\log h(t) = a(t) + 0.039 \text{ START} - 1.73 \text{ TYPE} + 0.028 \text{ START*TYPE}$$
$$\quad\quad\quad\quad\quad (0.007) \quad\quad\quad (0.62) \quad\quad\quad (0.016)$$

so there is a lower hazard for institution (TYPE=1), and some evidence for an interaction between start age and preschool type, such that the hazard for START is greater for TYPE=1.

# References

Agresti A (1988): A model for agreement between ratings on an ordinal scale. *Biometrics*, **44**, 539–548.
Aitkin M, Anderson D and Hinde J (1981): Statistical modelling of data on teaching styles (with discussion). *Journal of the Royal Statistical Society*, A, **144**, 148–61.
Aitkin M, Anderson D, Francis B and Hinde J (1989): *Statistical modelling in GLIM*. Oxford: Clarendon Press.
Armitage P and Berry G (1987): *Statistical Methods in Medical Research* (2nd edn). Oxford: Blackwell.
Barnett V (1991): *Sample Survey Principles and Methods*. London: Edward Arnold.
Bayley N (1949): Consistency and variability in the growth of intelligence from birth to eighteen years. *Journal of Genetic Psychology*, **75**, 165–96.
Bennett N (1976): *Teaching Styles and Pupil Progress*. London: OpenBooks.
Blalock HM (1979): *Social Statistics* (3rd edn). New York: McGraw-Hill.
Blossfeld H-P, Hamerle A and Mayer KU (1989): *Event History Analysis*. Lawrence Erlbaum: Hillsdale, NJ.
Bock RD (1979): Univariate and multivariate analysis of variance in time-structured data. In JR Nesselroade and PB Baltes (eds) *Longitudinal Research in the Study of Behavior and Development*. New York: Academic Press.
Brannen J and Moss P (1991): *Managing Mothers*. London: Unwin Hyman.
Bryk AS, Raudenbush SW and Congdon RT Jr. (1996): *Hierarchical Linear and Nonlinear Modelling with the HLM/2L and HLM/3L Programs*. Chicago: Scientific Software International.
Cook TD and Campbell DT (1979): *Quasi-experimentation*. Chicago: Rand McNally.
Cox DR and Oakes D (1984): *Analysis of Survival Data*. London: Chapman & Hall.
Cox DR and Snell EJ (1981): *Applied Statistics: Principles and Examples*. London: Chapman & Hall.

# 156  References

Cuttance P and Ecob R (eds) (1987): *Structural Modelling by Example.* Cambridge: Cambridge University Press.

Dale A, Arber S and Procter M (1988): *Doing Secondary Analysis.* London: Allen and Unwin.

Dale A and Davies RB (eds) (1994): *Analyzing Social and Political Change.* London: Sage.

de Gruijter DNM and van der Kamp LJTh (1991): Generalizability theory. In RK Hambleton and JN Zaal (eds) *Advances in Educational and Psychological Testing.* Boston: Kluwer.

Diggle P, Liang K-Y and Zeger SL (1994): *Analysis of Longitudinal Data.* Oxford: Clarendon Press.

Duncan C, Jones K and Moon G (1995): Blood pressure, age and gender. In G Woodhouse (ed) *A Guide to MLn for New Users.* London: Institute of Education, University of London.

Duncan OD (1984): *Notes on Social Measurement, Historical and Critical.* New York: Russell Sage Foundation.

EGRET (1993): *Reference Manual: First Draft, Revision 3.* Seattle: Statistics and Epidemiology Research Corporation.

Fleiss JL (1981): *Statistical Methods for Rates and Proportions* (2nd edn). New York: John Wiley.

Francis B, Green M and Payne C (eds) (1993): *The GLIM system, Release 4 Manual.* Oxford: Clarendon Press.

Freedman D, Pisani R, Purves R and Adhikari A (1991): *Statistics* (2nd edn). New York: Norton.

Fuller WA (1991): Regression estimation in the presence of measurement error. In PP Biemer, RM Groves, LE Lyberg, NA Mathiowetz and S Sudman (eds) *Measurement Errors in Surveys.* New York: Wiley.

Gilbert N (1993): *Analyzing Tabular Data: Loglinear and Logistic Models for Social Researchers.* London: UCL Press.

Goldstein H (1995): *Multilevel Statistical Models* (2nd edn). London: Edward Arnold.

Goldstein H and Sammons P (1997): The effectiveness of secondary and junior schools on sixteen year examination performance: a cross-classified multilevel analysis. *School Effectiveness and School Improvement*, **8**, 219–30.

Goldstein H and Spiegelhalter DJ (1996): League tables and their limitations: statistical issues in comparisons of institutional performance (with discussion). *Journal of the Royal Statistical Society*, A, **159**, 385–444.

Hagenaars J. (1990): *Categorical Longitudinal Data.* Newbury Park, CA: Sage.

Hox JJ (1995): AMOS, EQS and LISREL for Windows: A comparative review. *Structural Equation Modeling*, **2**, 79–91.

Jones K (1996): Review of HLM4 for Windows. *Multilevel Modelling Newsletter*, **8**, 3–6.

Kratochwill TR (ed.) (1978): *Single Subject Research: Strategies for Evaluating Change.* Orlando: Academic Press.

Kreft IGG, de Leeuw JB and Aiken LS (1995): The effect of different forms of centering in hierarchical linear models. *Multivariate Behavioral Research*, **30**, 1–21.

Langford I and Lewis T (1998): Outliers in multilevel data. *Journal of the Royal Statistical Society*, A, 161 (forthcoming).

Little RJA and Schenker N (1995): Missing data. In G Arminger, CC Clogg and ME Sobel (eds) *Handbook of Statistical Modelling for the Social and Behavioral Sciences*. New York: Plenum Press.

Lord FM and Novick MR (1968): *Statistical Theories of Mental Test Scores*. Reading, MA: Addison-Wesley.

McCullagh P and Nelder JA (1989) *Generalized Linear Models* (2nd edn). London: Chapman & Hall.

Manly BFJ (1986) *Multivariate Statistical Methods: A Primer*. London: Chapman & Hall.

Marsh C (1988): *Exploring Data*. London: Polity Press.

Paterson L (1995): Entry to university by school leavers. In G Woodhouse (ed.) *A Guide to MLn for New Users*. London: Institute of Education, University of London.

Plewis I (1981): A comparison of approaches to the analysis of longitudinal categoric data. *British Journal of Mathematical and Statistical Psychology*, **34**, 118–23.

Plewis I (1985): *Analysing Change*. Chichester: Wiley.

Plewis I, Mooney A and Creeser R (1990): Time on educational activities at home and educational progress in infant school. *British Journal of Educational Psychology*, **60**, 330–37.

Plewis I (1991a): Using multilevel models to link educational progress with curriculum coverage. In SW Raudenbush and J Willms (eds) *Schools, Classrooms and Pupils: International Studies of Schooling from a Multilevel Perspective*. San Diego: Academic Press.

Plewis I (1991b): Pupils' progress in reading and mathematics during primary school: Associations with ethnic group and sex. *Educational Research*, **33**, 133–40.

Plewis I (1995): Reading progress. In G Woodhouse (ed.) *A Guide to MLn for New Users*. London: Institute of Education, University of London.

Plewis I (1996a): Young children at school: Inequalities and the National Curriculum. In B Bernstein and J Brannen (eds) *Children, Research and Policy*. London: Taylor & Francis.

Plewis I (1996b): Statistical methods for understanding cognitive growth: a review, a synthesis and an application. *British Journal of Mathematical and Statistical Psychology*, **67**, 25–42.

Plewis I (1997): Inferences about teacher expectations from national assessment at Key Stage One. *British Journal of Educational Psychology*, **67**, 235–47.

Plewis I and Hurry J (1997): A multilevel perspective on the design and analysis of intervention studies. *Educational Research and Evaluation* (forthcoming).

Plewis I and Veltman M (1996): Opportunity to learn maths at Key Stage One: Changes in curriculum coverage 1984–1993. *Research Papers in Education*, **11**, 201–18.

Radical Statistics Education Group (1982): *Reading Between the Numbers*. London: BSSRS Publications.

Rasbash J and Woodhouse G (1995): *MLn Command Reference*. London: Institute of Education, University of London.

Rose D and Sullivan O (1996): *Introducing Data Analysis for Social Scientists*. Milton Keynes: Open University Press.

Schnell D, Park HJ and Fuller WA (1988): *EV CARP Manual*. Ames: Iowa State University.

Sewell WH and Shah VA (1968): Social class, parental encouragement and educational aspirations. *American Journal of Sociology*, **73**, 559–72.

Singer JD and Willett JB (1993): It's about time: using discrete-time survival analysis to study duration and the timing of events. *Journal of Educational Statistics*, **18**, 155–96.

Spiegelhalter DJ, Thomas A, Best NG and Gilks WR (1995): *BUGS Manual and Examples: Version 0.50*. Cambridge: Medical Research Council.

Sylva K, Hurry J and Plewis I (1995): *The Effectiveness of Reading Recovery and Phonological Training for Children with Reading Problems*. London: Schools Curriculum and Assessment Authority.

Tizard B, Blatchford P, Burke J, Farquhar C and Plewis I (1988): *Young Children at School in the Inner City*. Hove: Lawrence Erlbaum.

Tuma N (1994): Event history analysis. In A Dale and RB Davies (eds) *Analyzing Social and Political Change*. London: Sage.

Weisberg S (1980): *Applied Linear Regression*. New York: Wiley.

Woodhouse G, Yang M, Goldstein H and Rasbash J (1996): Adjusting for measurement error in multilevel analysis. *Journal of the Royal Statistical Society*, A, **159**, 201–12

Wright D (1996): Review of MLn 1.0a. *British Journal of Mathematical and Statistical Psychology*, **49**, 397–98.

# Author Index

Adhikari, A. 9, 31, 156
Agresti, A. 107, 155
Aiken, L.S. 50, 156
Aitkin, M. 34, 35, 44, 118, 127, 131, 141, 155
Anderson, D. 34, 35, 44, 118, 127, 131, 141, 155
Arber, S. 8, 156
Armitage, P. 81, 155

Barnett, V. 8, 155
Bayley, N. 58, 155
Bennett, N. 34, 155
Berry, G. 81, 155
Best, N.G. 146, 158
Blalock, H.M. 80, 155
Blatchford, P. 57, 78, 93, 158
Blossfeld, H.-P. 131, 155
Bock, R.D. 57, 155
Brannen, J. 115, 155
Bryk, A.S. 53, 146, 155
Burke, J. 57, 78, 93, 158

Campbell, D.T. 7, 155
Congdon, R.T. Jr. 53, 146, 155
Cook, T.D. 7, 155
Cox, D.R. 5, 85, 124, 131, 155
Creeser, R. 87, 157
Cuttance, P. 3, 156

Dale, A. 8, 156
Davies, R.B. 8, 156
de Gruijter, D.N.M. 138, 156
de Leeuw, J.B. 50, 156
Diggle, P. 145, 146, 156
Duncan, C. 144, 156
Duncan, O.D. 6, 156

Ecob, R. 3, 156

Farquhar, C. 57, 78, 93, 158
Fleiss, J.L. 106, 156
Francis, B. 44, 118, 127, 131, 155, 156
Freedman, D. 9, 31, 156
Fuller, W.A. 140, 156, 158

Gilbert, N. 95, 156
Gilks, W.R. 146, 158
Goldstein, H. 36, 42, 52, 67, 131, 144, 146, 156, 158
Green, M. 93, 156

Hagenaars, J. 141, 156
Hamerle, A. 131, 155
Hinde, J. 34, 35, 44, 118, 127, 131, 141, 155
Hox, J.J. 140, 156
Hurry, J. 53, 54, 157, 158

Jones, K. 53, 144, 156

Kratochwill, T.R. 8, 156
Kreft, I.G.G. 50, 156

Langford, I. 41, 156
Lewis, T. 41, 156
Liang, K.-Y. 145, 146, 156
Little, R.J.A. 30, 72, 157
Lord, F.M. 138, 157

McCullagh, P. 83, 95, 157
Manly, B.F.J. 3, 157
Marsh, C. 4, 77, 157
Mayer, K.U. 131, 155
Moon, G. 144, 156
Mooney, A. 87, 157
Moss, P. 115, 155

Nelder, J.A. 83, 95, 157

## Author index

Novick, M.R. 138, 157

Oakes, D. 124, 131, 155

Park, H.J. 140, 158
Paterson, L. 146, 157
Payne, C. 93, 156
Pisani, R. 9, 31, 156
Plewis, I. 8, 13, 24, 35, 50, 53, 54, 57, 59, 74, 75, 78, 87, 93, 105, 106, 107, 111, 144, 157, 158
Procter, M. 8, 156
Purves, R. 9, 31, 156

Radical Statistics Education Group 34, 157
Rasbash, J. 36, 53, 146, 157, 158
Raudenbush, S.W. 53, 146, 155
Rose, D. 95, 157

Sammons, P. 144, 156
Schenker, N. 30, 72, 157
Schnell, D. 140, 158
Sewell, W.H. 85, 158

Shah, V.A. 85, 158
Singer, J.D. 128, 158
Snell, E.J. 5, 85, 155
Spiegelhalter, D.J. 42, 146, 156, 158
Sullivan, O. 95, 157
Sylva, K. 54, 158

Thomas, A. 146, 158
Tizard, B. 57, 78, 93, 158
Tuma, N. 131, 158

van der Kamp, L.J.Th. 138, 156
Veltman, M. 13, 144, 157

Weisberg, S. 12, 15, 24, 30, 158
Willett, J.B. 128, 158
Woodhouse, G. 36, 53, 146, 157, 158
Wright, D. 53, 158

Yang, M. 146, 158

Zeger, S.L. 145, 146, 156

# Subject Index

accelerated life models 120
agreement
    chance 90
    measure of 91
    modelling 105–8
analysis of covariance (ANCOVA) model 27
analysis of variance (ANOVA) 20
    table 20, 22
arbitrary scales 6
    in growth curve modelling 61, 72
association 80, 81, 90, 94, 96–8, 105
    modelling 94–9
Attainment Targets (ATs) 144
attenuation effect 139
automatic selection of explanatory variables 29–30

balanced data 55
beta coefficient 14
binary data 6
binary responses, modelling 77–92
binary variables
    cross-tabulation 78
    repeated measurement 89
Binomial distribution 82, 87, 145, 146
borrowed information 38
BUGS 146

categorical data, unordered 6
categorical explanatory variables
    multiple regression with 24–6
    simple regression 18–20
categorical responses
    modelling 93–111

multilevel modelling 145–6
    order representation in 100
CATMOD procedure 101, 102, 104, 105, 108
causal explanation 3
causal models 7
causal relationship 12
censoring 117–18, 120, 122, 125, 131
centering 49–50, 60
change(s)
    causes 8
    in correlations with changes in scale 73
    in relative growth 73
    inferences 8
    modelling 103–5
    of scale 73
chi-square distribution 80, 81, 89, 104
chi-square statistic 104
chi-square value 84
coefficient of association 81
coefficient of variation (c.v.) 72
collinearity 52
competing risks models 115
complementary log-log function 128
complex variation at level one 67–9
conditional approach 74
conditional model 105
conditional probability 77
confidence intervals 38, 41, 79, 119
contextual variables 49
contingency tables 77, 80–1, 89, 93–111
continuation odds 100, 102
    analysis 109
continuation ratios 100
continuous data 5

## Subject index

continuous explanatory variables
    multiple regression with 20–4
    simple regression 11–16
continuous responses 11–32, 82, 145, 146
    multilevel models for 33–54
    repeated measures of 55–75
continuous time, event history data in 120–6
continuous variables 141
Cook's $D$ 24, 88
correlation coefficients 15, 139
correlations 71, 73
covariances 61, 71
covariates 2
Cox proportional hazards (PH) model 121, 123, 125, 130
criterion-referenced tests 74
cross-classified multilevel models 142–4
cross-classified structure 143
cross-level interaction 48, 51
cross-sectional surveys 7
cumulative hazard function 117, 119–20
cumulative logits 100, 101, 105
cumulative marginal probability 105
cumulative probability 100, 116

degrees of freedom (df) 20, 44, 49, 80, 81, 84, 86, 95, 96, 104
dependent samples 88–91
dependent variable 2
deviance 87, 88, 97, 98
discrete hazards 129
discrete time, event history data in 126–9
discrete time analysis, GLIM code for 135
discrete time hazard 127
distribution function 116
dummy variables 18–19, 20, 35, 88, 95, 121
duration 113, 120, 122, 124, 127
duration dependence 120

ecological fallacy 30–1
education as a career 113–36
educational data
    hierarchical nature of 33–5
    types 5–6
educational research designs 6–8
EGRET 146
episodes 115–16, 118, 142
    more than one 129–31
EVCARP 140
event history analysis 113–36
    dictionary of terms 132–4
    fundamental concepts 116–18
    measurement errors in 142
event history data 114–16
    in continuous time 120–6
    in discrete time 126–9
explanatory variables 2, 82, 84–8, 96, 120, 123, 127, 128, 140, 142, 144
    automatic selection 29–30
    principal components 30
exponential distribution 120
exponential models 125, 130
extra variation 145

factors 2
failure time 116
fitted counts and adjusted residuals 99
fixed effects 35, 47–50, 66, 72
    regression model 37
fixed scales 74
four-level structure 55

generalizability theory 138
generalized estimating equations (GEEs) 145
generalized linear model 83
GLIM code
    for discrete time analysis 135
    for Weibull analysis 134–5
GLIM4 125, 126, 128
goodness of fit 99, 145
growth rates 59–61
growth variability 64–7

hazard function 117, 121, 129
hazard models 114, 124, 125
hazards 117, 120
    *see also* proportional hazards
heaping 142
heteroscedasticity 15, 28, 73
hierarchical nature of educational data 33–5
hierarchical nature of repeated measures data 55–9
hierarchical (or nested) structures 34, 55, 143, 145
histograms
    curriculum coverage 13
    mathematics attainment 13
    standardized classroom residuals 37, 42
    standardized intercept residuals 45
    standardized pupil intercept residuals 50
    standardized pupil quadratic residuals 50
    standardized pupil slope residuals 50
    standardized school intercept residuals 64
    standardized school residuals 41
    standardized school slope residuals 45, 64
HLM 53, 146
homoscedasticity 15, 28

## Subject index

ignorable missingness 69
independence hypothesis 80, 95
independence model 95, 97
independence test 84
intake differences 40
integrated hazard 117
interactions 98, 99, 103
inter-rater reliability studies 89–90
interval scale 5
intervention studies, multilevel models for 50–3
intra class (or intra unit correlation) 36

joint probability 79

Kaplan-Meier method 118
kappa 91, 105–7
Kendall's tau-$c$ 81, 94
Key Stage One 144

latent class models 140
latent variable models 140
latent variables 138
league tables 40
least squares methods 12, 21, 120, 122
left censoring 117–18
level one complex variation 67–9
level one units 55–6
level two units 55–6
life table 116
linear by linear association 96
linear function 3
linear growth 60–1
　　model 60
linear transformation 16
log duration 122
　　model 123, 125
log-linear models 107, 108
　　for modelling association 94–9
log normal distribution 131
log transformation 95
logistic (or logit) link 82
logistic regression 77, 81–4
　　with more than one explanatory variable 84–8
longitudinal datasets 55
longitudinal designs 7, 56
longitudinal study 57

McNemar test 89, 103
main effects model 25
manifest variables 138
Mantel-Haenszel test 81, 84, 94
marginal homogeneity 103, 104
　　analysis 110

marginal probability 77, 89, 105
Markov models 114
maximum likelihood method 95
measurement bias 138
measurement errors 137–42
　　and true score, scatterplot 138
　　in event history analysis 142
　　model 141
　　variance 139
measurement model 138, 139
measurement scales 5–6
misclassification 140
missing at random 69–71
missing completely at random 69–70
missing data 30, 69–72
　　classes 69
missing non-ignorable data 70
MLn 36, 53, 59, 128, 146
model checking 4, 15, 22–4, 44, 86–7
model criticism 4, 15
monotone pattern of missing data 69
multilevel approach 55
multilevel growth curve models 59–64, 73, 74
multilevel models 5, 131
　　categorical data 145
　　cross-classified 142–4
　　extensions 142–4
　　for continuous responses 33–54
　　for intervention studies 50–3
　　for school and teacher effectiveness 38–42
　　multivariate 144
　　software 53
multiple episodes 131
multiple regression
　　with categorical explanatory variables 24–6
　　with continuous explanatory variables 20–4
　　with mixture of categorical and continuous explanatory variables 26–8
multiplicative model 95
multivariate analysis of variance (MANOVA) 57
multivariate logits 100
multivariate multilevel models 144
multivariate statistical models 3
multi-way tables 87, 96

nested models 99
nested structures 34, 143, 145
　　with two levels 55
non-ignorable missing data 69, 70
non-linear components 87

## Subject index

Normal distribution 3, 17, 28, 36, 37, 41, 43, 48, 60, 61, 82, 83, 88, 122
Normality
   assumption 87
   departure from 16
Normalizing transformation 16

observational studies 7
observed scores 138–9
odds 78, 82, 84, 100
odds ratio 78
one way analysis of variance 20
ordered categorical data 6
ordered categorical variables, representations of 100
ordered responses, modelling 101–3
ordered scales 5–6
ordering 105
ordinary least squares (OLS) estimates 40–1
outcome 2
overdispersion 145

partitioning variability by level 35–8
Pearson (or product moment) correlation 12
person epochs 127
person interval 127
Poisson distribution 94, 131, 145
Poisson errors 94
population averaged models 145
population probabilities 83
populations with structure 33–54
prediction 30
predictor variables 2
preschool episodes 120, 121, 123, 124
preschool experiences 113–16
preschool history 115
principal components of explanatory variables 30
probability density function 117
probability models 3
probability of failure 83
probability of success 83
product moment (or Pearson) correlation 12
proportional hazards 124
   models 120, 128
proportional odds analysis 108
proportional odds model 101, 128
proportionality assumption 124
$p$-value 81, 94

Q-Q plots 17, 61
   pupil intercept residuals 51
   pupil quadratic residuals 51
   pupil slope residuals 51
   school intercept residuals 65
   school slope residuals 65
   standardized residuals 87
   studentized residuals 122
   variance standardized residuals 126
quasi-independence model 96

random effects 35, 38, 39, 72
   models 35
random intercepts 42–7, 74
   model 50
random parameters 44, 60, 61
random slopes 42–7
randomized experiments 7
ratio scales 5
regression assumptions 28–9
regression coefficients 14, 19–21, 28, 35, 79, 139
regression estimates 140
regression models 14, 19, 29
   correcting 137–40
relative odds 78–9, 84
reliability
   definition 139
   estimates 140
repeated measures of continuous reponses 55–75
repeated measures data, hierarchical nature of 55–9
residual chi-square 105
residual deviance 85, 86, 96, 98
residuals 12
   adjusted (or variance stabilized) 87, 99, 126
   checking 60
   homoscedastic 28
   nature of 28
   scatterplot of 15, 17, 23
   simple regression model with 122
   studentized 15
response distribution 3
response variables 2, 120
right censoring 117–18
risk set 116, 118

sampling errors 79
SAS code 53, 101, 102, 104, 108–11
saturated model 84
scaling parameter 121
scatterplots
   intercept by slope residuals 46
   LOGMATH by ZMATH1 22
   mathematics attainment by curriculum coverage 14
   measurement error and true score 138

## Subject index

scatterplots *(cont.)*
    reading 40
    residuals 15, 17, 23
    school intercept and slope residuals 65
    standardized level one residuals 46
    standardized level one residuals by age 68, 69
    studentized residuals 124
secondary analysis 8
sensitivity 88
shrunken estimates 37
simple regression 46
    model 122, 139
    with categorical explanatory variable 18–20
    with continuous explanatory variable 11–16
Simpson's Paradox 85, 98
single subject designs 8
single unrepeated events 114
specificity 88
SPSS 57, 81, 83, 88, 91, 94, 95, 118, 124, 125
square root transformation 39
square tables
    generation 89
    with marginal homogeneity but not symmetry 103
standard assessment task (SAT) 106, 107, 108
standard deviations (SDs) 24, 58, 60, 70
standard errors (s.e.) 14, 15, 19, 35, 40, 52, 72, 79, 80, 83, 91, 94, 122, 139, 140
statistical inference 8–9
statistical models 1–5
    as conceptual tools 1
    strengths of 2
statistical significance 20
step-wise procedures 29
stochastic processes 114
structural equation modelling 3, 140
structured categorical data modelling 145–6
subject specific models 145
survey research 7
survival analysis 114
survival curves 116–17, 121
    construction 118–20

survival probability 100, 116
survival times 118, 119
survivor functions 116, 119
symmetry
    hypothesis 104
    McNemar test for 103

tau-$c$ 81, 94
three-level models 38, 40
three-way interaction 98
time dummies 128, 129
time series designs 8
transforming variables 16–18
transitions 113, 130
treatment variable 7
true scores 138–9
two by two tables 24, 77–80
two-level logistic regression model 145
two-level models 59, 71
    with random intercepts 39
    with random intercepts and random slopes 43
    with variance components 131
two-way tables 93–5

unconditional change modelling 110
unconditional model 105
unit specific models 145
unordered categorical data 6
unsaturated model 84

variability 1
variability in intercepts and slopes 47–50
variation *see* variability

Weibull analysis, GLIM code for 134–5
Weibull distributions 120, 121, 125
Weibull hazards 121
Weibull models 125, 126, 129, 130
Weibull survivor function 121
weighted kappa 106

Yates' correction 80
Yule's $Q$ 79

## OWNERSHIP OF COPYRIGHT

The copyright in the SOFTWARE and all copies of the SOFTWARE is owned by the author and is protected by United Kingdom copyright laws and international treaty provisions. As the Licensee, you merely own the magneticmedia or other physical media on which the SOFTWARE is recorded. You may take one copy of the program comprised in the SOFTWARE solely for backup or archival purposes. That copy must have the label placed on the magnetic media showing the program name, copyright and trademark designation in the same form as it appears on the original SOFTWARE. You may not copy the related printed materials.